Evidence-Based Weight Management

By

Milan Toma, PhD, SMIEEE

2025

Published by: Dawning Research Press
Contact: admin@dawningresearch.org
Website: www.dawningresearch.org

ISBN Number

Paperback: 979-8-9998324-6-7

All information has been carefully verified, and sources are cited throughout the text. The content reflects the author's commitment to evidence-based research and is intended for educational and informational purposes. The views expressed are those of the author and do not necessarily reflect those of any affiliated institutions. While every effort has been made to ensure the accuracy of the information presented, errors or omissions may still occur. If you spot a serious inaccuracy, we encourage you to contact us so that we can make corrections in the next edition. The author and publisher assume no responsibility for errors or omissions, or for damages resulting from the use of the information contained herein.

First Edition

Contents

Chapter 1

Why Do We Gain Weight?

Understanding why we gain weight requires careful interpretation of scientific evidence derived from epidemiological research and large-scale cohort analyses. The following sections synthesize the results of such research, highlighting patterns and associations found across extensive datasets and diverse cohorts. It is vital to emphasize that these findings represent general trends and statistical relationships; they do not, and cannot, explain or predict the experiences of every single person. Nutrition and dieting are extraordinarily complex sciences, and individual variability is the norm, not the exception. The analyses and conclusions presented here should therefore be viewed as informative for populations, not as one-size-fits-all explanations or prescriptions.

1.1 The Biology of Appetite Regulation

For decades, appetite regulation was viewed through a simplified lens: hunger drove eating, eating provided energy, and excess energy was stored as fat until needed. This mechanical model suggested that weight gain resulted from a straightforward imbalance between "calories in" and "calories out," with solutions focused primarily on conscious control of food intake and physical activity. However, advances in molecular biology and endocrinology have revealed that appetite and energy balance are governed by an extraordinarily complex network of hormonal signals, neural pathways, and metabolic feedback loops that operate largely below the level of conscious awareness. Far from being passive repositories for excess energy, fat cells actively communicate with the brain about the body's energy status through sophisticated chemical messengers. The stomach, intestines, liver, and other organs contribute their own regulatory signals, creating a biological control system of remarkable complexity. Understanding these mechanisms is vital for explaining why appetite regulation can fail, why some individuals are more susceptible to weight gain than others, and why simple approaches to weight management often prove inadequate. The following sections examine the key biological players in this system, beginning with the surprising discovery that fat tissue itself serves as a critical endocrine organ in the regulation of appetite and metabolism.

1.1.1 The Role of Fat Beyond Energy Storage

Fat tissue is often viewed solely as a passive reservoir for excess calories, but research has revealed that it is also a dynamic endocrine organ, producing hormones that play critical roles in regulating appetite, metabolism, and overall energy balance [1]. For example, in rare genetic conditions characterized by a near-total absence of fat tissue, individuals experience profound metabolic disturbances despite appearing extremely lean. These individuals lack the normal "padding" provided by fat, resulting in physical discomfort and a striking appearance of sunken features. More importantly, the absence of fat tissue deprives the body of essential hormonal signals, leading to severe metabolic consequences.

> **Key Point:** Fat tissue is not merely a storage depot for energy; it is an active endocrine organ whose absence leads to profound metabolic and physical consequences.

1.1.2 Lipodystrophy

In conditions such as generalized lipodystrophy, individuals are born with a normal amount of fat but rapidly lose it in early childhood, despite having an insatiable appetite. The lack of fat tissue means that excess calories have nowhere to be safely stored, resulting in their accumulation in organs such as the liver. This leads to liver enlargement, inflammation, and blood profiles that paradoxically resemble those seen in obesity (e.g., high levels of circulating fat and cholesterol) despite the absence of visible body fat [2]. The drive to eat is overwhelming, and attempts to restrict food intake can provoke extreme distress and behavioral changes, reflecting a state of perceived starvation.

> **Key Point:** The absence of fat tissue in lipodystrophy leads to severe metabolic disease, including fatty liver and abnormal blood lipids, despite extreme thinness and relentless hunger.

1.1.3 Leptin: A Hormone Linking Fat to the Brain

The biological basis for appetite regulation was illuminated by studies of a genetic mutant mouse, known as the OB (obese) mouse. This mouse carried a defect in a single gene, resulting in excessive eating and the accumulation of three times the normal body weight and five times as much fat as typical mice [3]. The mutation caused the mouse to overeat and become obese, demonstrating that a single gene could profoundly influence both appetite and body weight.

Through extensive genetic research, scientists identified that the affected gene was expressed only in fat cells and was responsible for producing a hormone. This hormone, later named leptin (from the Greek word for "thin"), was found to play a vital role in regulating appetite, metabolism, and other biological systems [4]. When leptin was administered to OB mice, their appetite normalized and their body weight returned to typical levels, demonstrating that leptin deficiency was the underlying cause of their obesity.

> **Key Point:** Leptin is a hormone produced by fat tissue that signals the brain about the status of energy stores, regulating appetite and metabolism.

1.1.4 Leptin Deficiency

Leptin acts as a messenger between fat tissue and the brain, particularly the hypothalamus, which governs hunger and satiety [5]. When fat stores are sufficient, leptin levels are high, signaling to the brain that energy reserves are adequate and suppressing appetite. When fat stores are depleted, leptin levels fall, triggering powerful drives to eat. In individuals who lack fat tissue, leptin is virtually absent, and the brain interprets this as a state of starvation, regardless of actual food intake or body composition. This leads to relentless hunger and metabolic dysfunction, including fatty liver and abnormal cholesterol levels.

Leptin replacement therapy in such individuals can rapidly reduce hunger and normalize eating behavior, often within days. While leptin does not restore lost fat tissue, it alleviates the sense of starvation and protects organs such as the liver from further damage.

> **Key Point:** Leptin deficiency causes the brain to perceive starvation, driving excessive hunger and metabolic disease; leptin replacement can normalize appetite and improve metabolic health, even without restoring fat tissue.

1.1.5   Leptin Resistance

In most individuals with obesity, leptin levels are not low but instead markedly elevated, reflecting the increased mass of fat tissue. Under normal circumstances, high leptin concentrations would signal to the brain (particularly the hypothalamus) that energy stores are abundant, suppressing appetite and promoting energy expenditure. However, in obesity, this feedback loop breaks down. Despite the abundance of leptin, the brain becomes less responsive to its signal, a phenomenon known as leptin resistance.

Leptin resistance is characterized by a paradox: the body has ample or even excessive energy reserves, yet the brain behaves as if it were in a state of deprivation. The hypothalamus, unable to "hear" the leptin signal, fails to adequately suppress hunger and continues to drive food-seeking behavior [6]. At the same time, energy expenditure may be reduced, as the body attempts to conserve what it mistakenly perceives as limited resources [7]. This mismatch between actual energy stores and the brain's perception of starvation creates a powerful biological drive to eat, even in the presence of obesity.

The origins of leptin resistance are complex and multifactorial. Chronic overnutrition, inflammation within the hypothalamus, and disruptions in cellular signaling pathways all contribute to the brain's diminished sensitivity to leptin. As a result, simply increasing leptin levels (either through further weight gain or by administering the hormone) does not restore normal appetite regulation in most people with obesity. Unlike individuals with true leptin deficiency, who respond dramatically to leptin replacement, those with leptin resistance experience little or no reduction in hunger or weight when given additional leptin.

This state of leptin resistance helps explain why weight loss is so difficult to achieve and maintain. As fat is lost and leptin levels fall, the brain's sensitivity to leptin does not automatically improve. Instead, the drop in leptin is interpreted as a threat to energy balance, triggering increased hunger and decreased energy expenditure. These adaptations persist long after weight loss, promoting weight regain and making sustained weight reduction a formidable challenge.

> **Key Point:** In obesity, high leptin levels fail to suppress hunger because the brain becomes resistant to leptin's signal. This leptin resistance causes the brain to act as if the body is starving, driving persistent hunger and reduced energy expenditure despite abundant fat stores.

1.1.6 Implications for Understanding Obesity

The discovery of leptin fundamentally changed the scientific understanding of obesity and appetite. It demonstrated that body fat is not simply a passive consequence of overeating, but an active participant in a feedback system that regulates energy balance. Obesity and leanness are not solely determined by willpower or lifestyle, but are influenced by complex hormonal signals that can override conscious control. In rare cases of leptin deficiency, restoring the missing hormone can correct the underlying drive to eat, highlighting the biological basis of appetite and the potential for targeted therapies in specific forms of metabolic disease.

> **Key Point:** The regulation of appetite and body weight is governed by hormonal signals from fat tissue, with leptin playing a central role; disruptions in this system can lead to profound metabolic and behavioral consequences.

1.1.7 Main Takeaway Points

o **Do recognize that fat tissue is an active endocrine organ.** Its absence leads to severe metabolic and behavioral disturbances, not just thinness.

o **Do understand the central role of leptin in appetite regulation.** Leptin signals the brain about energy stores, suppressing hunger when fat is sufficient and triggering hunger when it is low.

○ **Do appreciate that metabolic disease can occur in the absence of obesity.** Conditions like lipodystrophy demonstrate that fat distribution and hormonal signaling are as important as total fat mass.

○ **Do consider the biological basis of appetite and weight regulation.** Hormonal feedback systems, not just conscious choice, drive eating behavior and energy balance.

○ **Do note the therapeutic potential of hormone replacement in specific disorders.** Leptin therapy can normalize appetite and metabolic health in cases of leptin deficiency, even without restoring fat tissue.

> **Key Point:** Fat tissue and its hormone leptin are central to the regulation of appetite and metabolism. Disruption of this system, whether by genetic mutation or loss of fat, leads to severe metabolic disease and relentless hunger, underscoring the biological complexity of weight regulation.

1.2 Trends in Caloric and Nutrient Intake

A widely held belief is that the rise in obesity in the United States since the 1970s must be explained by a corresponding increase in calorie consumption or by shifts in macronutrient intake, such as more sugar or vegetable oil. This narrative is often supported by simple correlations between food supply data and obesity rates, or by anecdotal observations. However, a careful review of the evidence, including recent large-scale studies and methodological critiques, reveals a more complex and nuanced picture.

1.2.1 Correlation versus Causation

It is tempting to attribute the rise in obesity to a dietary factor that closely parallels obesity rates on a graph; whether that's vegetable oils, sugar, or total calories. However, if we relied on nothing more than two lines moving together, we could just as easily conclude that eating more cheese increases your chances of dying from getting entangled in your bedsheets while sleeping, since those two statistics also track closely on a graph. Such examples highlight why correlation does not equal causation and why spurious associations are so common and misleading. More importantly, the reliability of the underlying data must be scrutinized before drawing any conclusions.

> **Key Point:** Correlation between dietary trends and obesity does not establish causation. Detailed analysis and high-quality data are vital before making policy or personal recommendations.

1.2.2 What Do National Surveys Show?

The National Health and Nutrition Examination Survey (NHANES) is the primary source of dietary intake data in the United States. Analyses of NHANES data have been widely cited to support the claim that average calorie intake has increased over time, particularly from the 1970s through the early 2000s. For example, among Mexican-American adults, mean energy intake rose from 1,983 kcal in 1982–1984 to 2,186 kcal in 1999–2006, with increases observed in both men and women [8]. Over the same period, the percentage of calories from fat and saturated fat decreased, while carbohydrate intake increased, and protein intake as a percentage of calories declined slightly. These trends mirror those seen in the general U.S. population and are consistent with public health recommendations to reduce fat intake during this period [8].

Despite these changes, the prevalence of obesity and diabetes increased substantially among Mexican-American adults and the broader U.S. population [8]. This suggests that simple macronutrient composition is not the sole driver of obesity trends, and that other factors (including total energy intake, physical activity, food environment, and underreporting) must be considered.

> **Key Point:** Shifts in macronutrient composition (e.g., lower fat, higher carbohydrate) have not prevented the rise in obesity, indicating that the causes are multifactorial and not explained by any single nutrient.

1.2.3 The Validity of Dietary Surveillance

However, these findings must be interpreted with caution. Decades of methodological research have demonstrated that self-reported dietary intake data (such as that collected in NHANES) are subject to substantial underreporting, particularly among individuals with overweight or obesity. Analyses comparing reported energy intake to what would be physiologically plausible for survival and maintenance reveal that the majority of respondents in U.S. nutrition surveys report calories far too low to sustain basic metabolic needs. For example, across nearly forty years of NHANES, 67% of women and 59% of men reported energy intakes that were not physically plausible, with the extent of underreporting increasing sharply with higher BMI [9]. In obese adults, the average underreporting exceeded 700–850 kcal per day, or up to 40% of estimated needs.

Furthermore, changes in survey methodology over time (such as the implementation of improved interview techniques and the inclusion of more weekend days) have introduced artificial increases in reported calorie intake, rather than reflecting true changes in eating behavior [9]. The result is that much of the apparent increase in average calorie intake seen in national nutrition surveys may be explained by these methodological changes, not by real shifts in consumption.

Self-reported food surveys are also vulnerable to selective misreporting. People are especially likely to underreport foods perceived as unhealthy, such as those high in fat or sugar; a tendency that has grown as public awareness of dietary health risks has increased [9]. As a result, apparent trends in macronutrient intake may reflect shifting patterns of honesty or social desirability bias, rather than actual dietary change.

Finally, the process of converting food recall data into nutrient intake estimates relies on food composition databases that have changed over time and may not accurately reflect the nutrient content of all foods, especially those eaten outside the home or from commercial sources. This further complicates efforts to track true trends in energy and nutrient intake over decades [9].

> **Key Point:** Self-reported dietary intake data systematically and substantially underestimate actual calorie consumption, especially in people with overweight or obesity. Trends in reported calorie intake may reflect changes in survey methods or reporting bias, rather than real changes in what Americans eat.

1.2.4   Micronutrient Trends

Recent analyses of NHANES data from 2003–2018 show that, despite decades of dietary guidance, U.S. adults continue to underconsume key micronutrients (vitamins A, C, D, E, and K; calcium; potassium; magnesium; and choline), even when dietary supplements are included [10]. Sodium remains overconsumed. While some progress has been made for certain nutrients, overall nutrient intake and the percentage of the population meeting recommendations have changed little over the past 15 years. Notably, dietary supplements now provide a substantial share of intake for several vitamins and minerals, but even with supplements, many adults do not meet recommended intakes [10].

The use of dietary supplements has itself increased over time. Between 2007 and 2018, the prevalence of any supplement use in U.S. adults rose from 54% to 61%, with increases especially notable among men, non-Hispanic Black and Hispanic adults, and those with lower incomes [11]. The use of single-nutrient supplements (e.g., vitamin D, magnesium, vitamin B-12) increased, while multivitamin-mineral use declined. Among children, supplement use remained stable, except for an increase among food-insecure children [11].

> **Key Point:** Despite increased supplement use, most U.S. adults still do not meet recommendations for several key micronutrients, and overall dietary quality remains suboptimal.

1.2.5 Main Takeaway Points

- **Do not assume that simple trends in self-reported calorie or macronutrient intake explain the rise in obesity.** The data are confounded by underreporting and methodological changes [9].

- **Do not rely on visual correlations between dietary factors and obesity rates.** Correlation does not establish causation, and spurious associations are common.

- **Do recognize the limitations of dietary surveillance data.** Most self-reported intake data are not physiologically plausible, especially in people with obesity [9].

- **Do focus on objective measures and high-quality research.** When possible, use studies that validate dietary intake with biomarkers or doubly labeled water (see the next section).

- **Do not oversimplify the causes of obesity.** The epidemic is multifactorial, involving complex interactions between diet, physical activity, environment, and biology [8,10].

○ **Do not assume that increased supplement use has solved nutrient inadequacy.** Even with supplements, many Americans do not meet recommendations for key micronutrients [10, 11].

> **Key Point:** The rise in obesity cannot be explained by simple trends in self-reported calorie or macronutrient intake. Underreporting, methodological changes, persistent micronutrient shortfalls, and multifactorial causes demand caution in interpreting dietary surveillance data and forming public health recommendations.

1.3 Underreporting of Food Intake

A persistent challenge in nutrition research is the accuracy of self-reported dietary intake. This issue is especially pronounced among individuals with overweight or obesity, who tend to under-report their food and calorie consumption to a much greater extent than lean individuals [9]. In fact, the problem of underreporting among people with obesity is so substantial and well-documented that it warrants its own focused discussion. Understanding both the magnitude and the implications of this underreporting is vital for interpreting national dietary surveillance data and for making sense of the relationship between reported intake and obesity trends.

1.3.1 Systematic Underreporting in Obesity

Numerous studies have shown that people, on average, under-report their food intake when asked to recall what they eat. However, this tendency is not uniform across the population. Individuals with higher body mass index (BMI) are much more likely to under-report, both in terms of total calories and in specific "unhealthy" foods such as those high in fat or sugar [9]. In the NHANES surveys, for example, the majority of obese respondents reported calorie intakes that were not even sufficient to maintain basic metabolic functions, let alone support their actual body weight [9]. This pattern is so consistent that it has become a defining feature of self-reported dietary data in obesity research.

> **Key Point:** Obese individuals systematically and substantially under-report their food and calorie intake, making self-reported dietary data unreliable for this group.

1.3.2 The Role of Doubly Labeled Water

The gold standard for objectively measuring how many calories a person actually burns (and, in weight-stable individuals, consumes) is the doubly labeled water (DLW) method. This technique involves giving a person water in which both the hydrogen and oxygen atoms are replaced with rare, non-radioactive isotopes. After ingestion, these isotopes mix with the body's water and are gradually eliminated through urine, sweat, and breath. By tracking the rate at which the isotopes disappear from the body, scientists can precisely calculate the person's total energy expenditure over a period of days or weeks.

Because, in the long run, energy intake must match energy expenditure for weight to remain stable, DLW provides an objective benchmark against which to compare self-reported food intake. When researchers have used DLW to validate self-reported dietary data, the results have been striking: obese individuals under-report their calorie intake by an average of 700–850 kcal per day, or up to 40% of their actual needs [9]. In contrast, lean individuals tend to under-report by a smaller margin, typically 10–20%. The degree of underreporting increases with BMI, and is especially pronounced for foods perceived as unhealthy.

> **Key Point:** Doubly labeled water studies provide an objective measure of calorie intake, and consistently show that obese individuals under-report their intake far more than lean individuals.

1.3.3 Evidence from National Surveys

Analysis of nearly four decades of NHANES data reveals the extent of this problem. Across all survey years, 67% of women and 59% of men reported energy intakes that were not physiologically plausible, i.e., too low to sustain life at their measured body size [9]. Among obese adults, the proportion of implausible reporters was even higher, and the average underreporting exceeded 700–850 kcal per day. In some survey cycles, fewer than 1 in 4 obese women reported plausible calorie intakes. These findings are not unique to the United States; similar patterns have been observed in other countries and in a variety of study designs [9].

1.3.4   Why Does Underreporting Happen?

Several factors contribute to the tendency of obese individuals to under-report their intake. Social desirability bias (the desire to give answers that are viewed favorably by others) plays a major role, especially as public awareness of the health risks of obesity and certain foods has increased. People may also genuinely forget or misestimate portion sizes, especially for snacks, beverages, or foods eaten outside the home. Selective underreporting is particularly common for foods high in fat or sugar, which are often labeled as "bad" in public health messaging [9].

1.3.5   Implications for Research and Policy

The systematic underreporting of food intake by obese individuals has profound implications for nutrition research and public health policy. It means that national dietary surveillance data, which rely on self-reported intake, cannot be used to accurately estimate true calorie consumption in the population, especially among those most at risk for obesity-related diseases. Apparent trends in calorie or nutrient intake may reflect changes in reporting behavior or survey methodology, rather than real changes in eating habits. This limitation must be acknowledged when interpreting studies that attempt to link dietary trends to the rise in obesity.

> **Key Point:** Because obese individuals under-report their intake to a much greater extent than lean individuals, self-reported dietary data cannot be used to accurately assess true calorie consumption or its relationship to obesity trends.

1.3.6   Main Takeaway Points

○ **Do not take self-reported calorie or food intake at face value, especially in studies of obesity.** The data are systematically biased and underestimate true intake, particularly among those with higher BMI [9].

○ **Do not assume that reported declines or increases in calorie intake reflect real changes in eating behavior.** Changes in survey methods or social desirability bias may be responsible for apparent trends.

○ **Do use objective measures, such as doubly labeled water, when possible, to validate dietary intake data.** These methods provide a more accurate picture of true energy consumption.

○ **Be cautious when interpreting associations between self-reported diet and health outcomes in obese populations.** The magnitude of underreporting can distort observed relationships.

○ **Recognize that underreporting is a major barrier to understanding the true drivers of obesity at the population level.** Policy and recommendations should be based on objective evidence whenever possible.

> **Key Point:** Obese individuals under-report their food intake far more than lean individuals, as revealed by objective studies using doubly labeled water. This systematic bias undermines the validity of self-reported dietary data in obesity research and public health surveillance.

1.4   Total and Basal Energy Expenditure

Understanding why we gain or lose weight requires a clear grasp of how the body uses energy: not just how much we eat, but how much we burn. The concepts of *Total Energy Expenditure* (TEE) and *Basal Energy Expenditure* (BEE, also known as Basal Metabolic Rate, BMR) lie at the heart of energy balance. Importantly, BEE/BMR is not a separate phenomenon from TEE, but rather its largest and most stable component. This section introduces their definitions, how they are measured, how they change with age, and why common beliefs about metabolism are often mistaken.

1.4.1   Defining Total and Basal Energy Expenditure

Total Energy Expenditure (TEE) is the sum of all calories burned by the body in a day. It consists of three main components: (1) **Basal Energy Expenditure (BEE/BMR)**, which is the energy needed to maintain basic bodily functions at rest; such as breathing, circulating blood, and maintaining body temperature; (2) the **Thermic Effect of Food (TEF)**, which is the energy required to digest, absorb, and process the nutrients in food; and (3) **Physical Activity Energy Expenditure (PAEE)**, which is the energy spent on all movement; including structured exercise and daily activities. Of these, BEE/BMR is the largest component, typically accounting for 60–70% of daily energy use in adults [12–14].

Basal Energy Expenditure (BEE/BMR) refers specifically to the calories burned at complete rest, in a fasted state, and at thermoneutral temperature. It is the "idling cost" of keeping you alive, and is strongly determined by your fat-free mass (especially organs and muscle), age, sex, and genetics [12,13]. Thus, BEE/BMR is not separate from TEE, but rather forms its foundation.

> **Key Point:** TEE is the sum of all calories burned daily, with BEE/BMR as its largest and most stable component. BEE/BMR reflects the energy needed for basic bodily functions at rest and is determined mainly by fat-free mass, age, sex, and genetics; not a separate process from TEE, but its foundation.

1.4.2 How TEE and BEE/BMR Are Measured

The most accurate way to measure TEE in free-living humans is the *doubly labeled water* (DLW) method, which involves drinking water containing stable isotopes of hydrogen and oxygen and then tracking their elimination through urine, sweat, and breath over several days [12–14]. This provides a precise, objective measure of average energy expenditure. Basal energy expenditure (BEE/BMR) is measured in the lab, typically via indirect calorimetry, which calculates energy use from measurements of oxygen consumption and carbon dioxide production while the subject is at rest [13, 14]. These objective, biomarker-based methods are vital for overcoming the well-documented limitations of self-reported dietary and activity data [9].

> **Key Point:** TEE is most accurately measured by doubly labeled water, while BEE/BMR is measured by indirect calorimetry in the lab. These objective, biomarker-based methods are essential for overcoming the limitations of self-reported dietary and activity data and provide a reliable foundation for understanding energy balance.

1.4.3 TEE and BEE/BMR Across the Lifespan

Recent large-scale studies using DLW and indirect calorimetry have fundamentally revised our understanding of how energy expenditure changes with aging. Adjusted for body size and composition, TEE and BMR peak in infancy, decline through childhood and adolescence, and then reach stable adult levels by around age 20. Contrary to popular belief, both TEE and BMR remain remarkably stable from ages 20 to 60, even during pregnancy. Only after age 60 does metabolism begin to decline, with TEE and BMR dropping by about 0.7% per year; by age 90, energy expenditure is roughly 25% lower than in middle age [12]. These patterns have been confirmed in diverse populations and even in highly active groups [12–14].

The age-related decline after 60 is largely due to loss of fat-free mass, especially muscle and high-metabolic-rate organs, as well as reductions in the metabolic rate of individual tissues [12,13,15]. Body composition, not just age, is a major determinant of metabolic needs; lean mass is the primary driver of BEE/BMR [12,13].

> **Key Point:** TEE and BEE/BMR peak in infancy, decline through childhood, and stabilize in adulthood from ages 20 to 60, only declining after age 60. The age-related decline in later life is mainly due to loss of lean mass and reduced tissue metabolic rates, making body composition (not just age) a key determinant of metabolic needs.

1.4.4   Common Myths

A widespread myth is that adult weight gain is an inevitable result of a dramatic "metabolic slowdown" in midlife. However, the best available evidence from doubly labeled water studies shows that, after accounting for body size and fat-free mass, metabolism is stable from ages 20 to 60 [12]. Instead, weight gain in adulthood is primarily driven by gradual lifestyle changes, i.e., decreases in physical activity, increases in calorie intake, and shifts in body composition (more fat, less muscle), rather than any sudden drop in metabolism. In later life, true metabolic decline occurs, but it is gradual and results from both loss of lean mass (including muscle and organs) and reductions in the metabolic rate of these tissues [12,14,15].

Another misconception is that "exercise boosts metabolism" enough to prevent weight gain. While physical activity is vital for health and increases TEE, its direct impact on energy balance is often smaller than expected, unless it is substantial and sustained, and unless it is not offset by reductions in other activities or increases in calorie intake [16]. Furthermore, the largest component of TEE, namely BMR/BEE, is determined mostly by organ and lean tissue mass, not by exercise per se.

For older adults, traditional equations (such as the Schofield equation) often overestimate energy needs, as they were derived from younger, more active populations. More recent equations (e.g., Mifflin, Ikeda, Livingston) provide closer approximations for those over age 80 [13]. This has important implications for clinical nutrition and public health recommendations.

Comprehending the components and measurement of energy expenditure is essential for interpreting caloric needs, setting realistic weight management goals, and designing effective interventions. It also helps explain why weight loss is often more difficult to maintain than to achieve: as body mass (especially lean mass) declines with weight loss, BMR/BEE drops, reducing total calorie needs and potentially promoting increases in fat mass unless intake and activity are carefully matched [13, 15].

> **Key Point:** Contrary to popular belief, adult weight gain is not caused by a dramatic metabolic slowdown in midlife. Metabolism remains stable from ages 20 to 60; weight gain is driven by lifestyle and body composition changes. Exercise increases TEE but does not substantially raise BMR/BEE, and traditional equations may overestimate energy needs in older adults.

1.4.5   Practical Implications

A clear understanding of TEE and BEE/BMR is vital for everyone interested in weight management, healthy aging, and nutritional adequacy. Accurate estimates of energy needs are essential for preventing both obesity and undernutrition, especially among older adults and those with changing body composition. Using up-to-date, validated equations ensures that dietary advice matches actual requirements [13]. Recognizing that metabolism is mostly stable until the 60s shifts the focus from blaming "slow metabolism" to addressing lifestyle, environment, and behavior.

Hence, TEE reflects all calories burned daily, and BMR/BEE is its largest, most stable component. Midlife weight gain is not due to a slowing metabolism, but to lifestyle and body composition changes. Objective methods (DLW, indirect calorimetry) and modern equations should be used to assess energy needs, especially in older adults. Lean mass is the primary driver of BMR/BEE, and while exercise is vital for health, its impact on energy expenditure is often less than expected unless accompanied by increases in total activity and careful dietary management [16].

> **Key Point:** Accurate understanding and measurement of TEE and BEE/BMR are vital for setting realistic weight management goals and preventing both obesity and undernutrition, especially in older adults. Modern, validated equations and objective methods should guide dietary advice, with a focus on body composition and lifestyle rather than blaming "slow metabolism."

1.4.6   Main Takeaway Points

- ○ **Do understand that TEE reflects all calories burned daily, and BMR/BEE is its largest, most stable component.**

- ○ **Do not blame "slowing metabolism" for midlife weight gain; TEE and BMR remain stable from 20–60 years of age.**

- ○ **Do recognize that accurate measurement (DLW, indirect calorimetry) is vital for understanding energy balance.**

- ○ **Do appreciate that body composition, not just age, determines energy needs.**

- ○ **Do use modern, validated equations for estimating energy requirements in older adults.**

- ○ **Do remember that physical activity increases TEE, but its effect on body weight depends on total energy balance and compensatory changes in other components.**

> **Key Point:** Total and basal energy expenditure form the foundation of energy balance. Contrary to popular belief, metabolism is stable in adulthood and declines gradually with advancing age, primarily due to loss of lean mass and reductions in tissue metabolic rates. Accurate measurement and understanding of these concepts are vital for realistic, effective weight management and dietary guidance [12–15].

1.5 Metabolic Aging

As briefly introduced in the previous section (1.4), the finding that metabolism remains remarkably stable from ages 20 to 60 is so significant (and so contrary to popular belief) that it merits its own section. The notion that our metabolism inevitably slows as we age, making weight gain unavoidable, is one of the most deeply ingrained assumptions about aging. This assumption has shaped not only popular understanding but also clinical practice, with both patients and healthcare providers frequently attributing midlife weight gain to "slowing metabolism."

The belief is so entrenched that it often serves as both explanation and excuse for the gradual weight accumulation that many adults experience in their thirties, forties, and fifties. However, recent research using gold-standard methods for measuring energy expenditure has fundamentally challenged this conventional wisdom, revealing a far more nuanced and surprising picture of how human metabolism actually changes across the lifespan Understanding these findings is vital not only for dispelling myths about aging and weight gain, but also for developing more effective strategies for lifelong weight management based on biological reality rather than popular misconceptions.

1.5.1 How Does Metabolism Change With Age?

A common belief is that weight gain in adulthood is largely the result of a slowing metabolism. However, recent large-scale studies using doubly labeled water (DLW, explained in more detail in section 1.3), i.e., the gold standard for measuring total energy expenditure (TEE), have fundamentally revised our understanding of how human metabolism changes across the lifespan.

In a global analysis of over 6,400 individuals aged 8 days to 95 years, four distinct metabolic life stages were identified [12]. After adjusting for body size and composition, TEE accelerates rapidly in infancy, peaks at about 1 year (nearly 50% higher than adult values), then steadily declines through childhood and adolescence, reaching adult levels by age 20. Contrary to popular belief, adjusted TEE and basal metabolic rate (BMR) remain remarkably stable from ages 20 to 60, even during pregnancy. Only after age 60 does metabolism begin to decline, with TEE and BMR dropping by about 0.7% per year; so that by age 90, energy expenditure is roughly 25%

lower than in middle age [12].

> **Key Point:** Metabolism does not slow during midlife. Instead, total and basal energy expenditure remain stable from ages 20 to 60, only declining after age 60 [12].

1.5.2   Weight Gain With Age

Despite stable metabolic rates in adulthood, body weight and fat mass tend to increase from the fourth decade of life onward, with average weight gain of 0.3 to 0.5 kg per year between ages 40 and 66 [17]. This weight gain is not uniform: older adults accumulate more central (visceral) fat, which is metabolically active and associated with increased risk of insulin resistance and cardiometabolic disease. At the same time, lean mass declines, a process known as sarcopenia. These changes in body composition alter metabolic function, as lean tissue is the primary determinant of energy requirements [17].

Resting energy expenditure (REE) decreases with age, but this decline is out of proportion to the loss of lean mass, suggesting additional age-related changes in tissue metabolism [12, 17]. The thermic effect of food (TEF) and energy expenditure from physical activity also decrease, largely due to reduced activity levels in older adults. However, studies show that even highly active older adults experience some decline in muscle mass and function, indicating a primary effect of aging beyond lifestyle factors [17].

> **Key Point:** Age-related weight gain is driven by increased fat mass (especially visceral fat) and loss of lean mass, rather than a midlife slowdown in metabolism [12, 17].

1.5.3   Nonlinear Changes Across the Lifespan

Recent advances in multi-omics profiling have revealed that the biological processes underlying aging and metabolic health do not change in a simple, linear fashion. In a longitudinal study of 108 adults aged 25 to 75, comprehensive profiling of transcripts, proteins, metabolites, cytokines, and microbiome composition showed that most molecular markers of aging change nonlinearly, with two major periods of dysregulation occurring around ages 44 and 60 [18]. These transitions correspond to waves of molecular and functional changes, including shifts in immune regulation, carbohydrate and lipid metabolism, and increased risk for cardiovascular, metabolic, and neurodegenerative diseases.

For example, the prevalence of cardiovascular disease in the U.S. rises from about 40% between ages 40 and 59 to 75% between 60 and 79, and 86% in those over 80 [18]. Similarly, markers of kidney function, glucose metabolism, and immune regulation show nonlinear changes, with sharp inflection points around age 60. These findings suggest that the risk of age-related diseases accelerates at specific points in the lifespan, rather than increasing steadily with age.

> **Key Point:** Most molecular and metabolic changes during aging are nonlinear, with major transitions around ages 44 and 60 that correspond to increased disease risk [18].

1.5.4 Implications for Weight Management

The stability of metabolism in adulthood means that midlife weight gain cannot be blamed on a slowing metabolic rate. Instead, weight gain is more likely due to changes in body composition, reduced physical activity, and complex, nonlinear shifts in molecular pathways that affect appetite, fat storage, and energy balance [12, 17, 18]. Interventions that preserve muscle mass, promote physical activity, and support metabolic health (such as resistance training, adequate protein intake, and possibly omega-3 fatty acids) are vital for preventing age-related weight gain and metabolic dysfunction [17].

Moreover, the identification of nonlinear molecular transitions suggests that there may be critical windows for intervention to prevent or delay the onset of age-related diseases. Understanding these inflection points could inform personalized strategies for maintaining healthspan and preventing obesity and its complications.

> **Key Point:** Weight gain in adulthood is not caused by a midlife metabolic slowdown, but by changes in body composition, activity, and nonlinear molecular shifts. Targeted interventions during key transitions may help prevent obesity and age-related diseases [12, 17, 18].

1.5.5   Main Takeaway Points

○ **Do not blame midlife weight gain on a slowing metabolism.**
Metabolic rate remains stable from ages 20 to 60, declining only in later
life [12].

○ **Do recognize the importance of body composition.** Age-related
increases in fat mass and loss of muscle drive changes in metabolic health
and energy needs [17].

○ **Do appreciate the nonlinear nature of aging.** Most molecular and
metabolic changes occur in waves, with major transitions around ages
44 and 60 [18].

○ **Do focus on maintaining muscle mass and physical activity.**
These are key to preventing age-related weight gain and metabolic
dysfunction [17].

○ **Do consider the timing of interventions.** Targeting critical periods
of molecular change may improve prevention of obesity and chronic
disease [18].

> **Key Point:** The causes of adult weight gain are multifacto-
> rial and nonlinear, involving stable metabolism until late life,
> shifts in body composition, and waves of molecular change that
> increase disease risk at specific ages. Effective prevention re-
> quires attention to muscle mass, activity, and the timing of
> interventions [12, 17, 18].

1.6   The Cycle of Weight Gain and Dieting

For millions of adults, the pattern is frustratingly familiar: gradual weight
gain over the years, followed by determined efforts to lose weight, temporary
success, and then a return to the previous weight or higher. This cycle of
weight gain and dieting has become so common that it is often accepted
as an inevitable part of aging. The typical explanation offered, and widely
believed, is that metabolism naturally slows with age, making weight gain
unavoidable and weight loss increasingly difficult with each passing decade.
However, this conventional wisdom, while intuitively appealing, is not supported by scientific evidence. The following sections examine what actually

drives weight cycling, challenge widespread misconceptions about metabolic changes with age, and reveal the complex biological and behavioral factors that make sustained weight management so challenging. Understanding these mechanisms is vital for developing realistic expectations and effective strategies for long-term weight control.

1.6.1 Beyond the Metabolism Myth

For many people, midlife brings gradual weight gain; often explained, both popularly and professionally, as the inevitable result of a "slowing metabolism." This belief has influenced generations of dieters and clinicians alike. However, recent large-scale studies using doubly labeled water and multi-omics profiling have overturned this assumption.

After adjusting for body size and composition, total and basal energy expenditure remain remarkably stable from ages 20 to 60 [12]. The much-feared midlife "metabolic slowdown" simply does not occur for most adults. Instead, metabolism only begins to decline after age 60, and even then at a gradual rate [12].

So why do so many of us gain weight in our 30s, 40s, and 50s? The answer lies in lifestyle changes that often come with work, family, and social responsibilities. As we age, we typically become more sedentary, spend more hours sitting at desks or in cars, have less time for exercise, and face increased demands that can displace physical activity and healthy eating. Simultaneously, social gatherings, work-related meals, and convenient, calorie-dense foods become more common. These factors combine to create a gradual energy imbalance, resulting in slow but persistent weight gain over the decades [17]. Biological pressures can also interact with these environmental and behavioral influences to establish a new, higher "steady state" weight [19].

> **Key Point:** Most weight gain in adulthood is not caused by a slowing metabolism, but by lifestyle changes that reduce physical activity and increase calorie intake [12, 17].

1.6.2 Each Weight Gain Makes Weight Loss Harder

Many people, upon noticing their weight creeping up, resolve to "get back" to their earlier, "happier" weight. However, attempts to lose weight (especially through rapid or restrictive diets) are often met with powerful biological resistance. When we lose weight, particularly through caloric restriction, our bodies interpret the calorie deficit as a threat to energy stores. This triggers adaptive responses: hunger increases, satiety signals weaken, and, most importantly, metabolism slows down to levels lower than would be expected for the new, lower body weight [17, 19]. This phenomenon, sometimes called "adaptive thermogenesis," is an evolved response to perceived starvation.

Extensive research now demonstrates that the biological response to weight loss is both persistent and multidimensional: reductions in leptin and insulin levels signal depleted energy stores to the brain, hypothalamic and neuroendocrine pathways shift to increase hunger and decrease satiety, and energy expenditure (including both resting and non-resting components) declines even after adjusting for reduced body mass [19]. These adaptations are particularly pronounced in individuals with obesity, and can persist long after weight loss, making further weight loss and maintenance increasingly difficult.

The brain, too, plays a central role. After weight gain, the brain comes to accept and defend the higher weight as the "new normal." When weight is lost, especially quickly, the brain and body coordinate to restore the lost weight, interpreting the deficit as a danger signal [17, 19]. As a result, people who lose weight often experience a persistent drop in resting metabolic rate, lower energy expenditure during activity, and stronger drives to eat, sometimes for years after weight loss [19].

> **Key Point:** During and after dieting to lose weight, the brain and body continue to defend the previous, higher weight, making further weight loss harder and promoting weight regain through metabolic and behavioral adaptations [17, 19].

1.6.3   The Vicious Cycle of Dieting and Weight Regain

If, after dieting, a person returns to their previous eating and activity patterns, they may regain weight even more rapidly than before. This is because their metabolism remains suppressed from the prior caloric deficit, while their calorie intake has returned to earlier levels [17, 19]. With each cycle of weight loss and regain, the body can become even more efficient at conserving energy, and the baseline metabolic rate may decrease further. This "yo-yo" pattern (known as weight cycling) makes it progressively harder to lose weight and easier to regain it, trapping many people in a frustrating cycle of dieting, rebound, and weight gain.

MacLean et al. [19] highlight that this "energy gap" (i.e., the difference between increased appetite and decreased energy expenditure) drives powerful physiological urges to eat in excess of one's new, lower requirements. Adaptive changes in adipose tissue, muscle, and liver, as well as persistent neuroendocrine and gut hormone changes, all conspire to promote weight regain. Notably, the authors propose that the homeostatic system in the body becomes "retuned" to defend a higher weight, so after weight loss, the biological drive to regain weight is even stronger than the original drive to gain it.

Multi-omics studies now reveal that these cycles are not just behavioral or psychological, but are accompanied by waves of biological changes in metabolism, inflammation, tissue structure, and disease risk, with major transitions occurring around ages 40–45 and 60–65 [18]. Each cycle of weight gain and loss may further dysregulate these molecular pathways, increasing the risk of metabolic and cardiovascular disease.

> **Key Point:** Repeated cycles of weight loss and regain (weight cycling) may further slow metabolism, disrupt biological pathways, and increase the risk of disease [17–19].

1.6.4 Prevention Is Easier Than Cure

Given these realities, the most effective strategy for lifelong weight management is to avoid significant weight gain in the first place. When major life changes occur (such as starting a demanding job, raising children, or experiencing stress) adjusting eating habits and maintaining physical activity are vital. By matching calorie intake to actual energy needs and staying active, you can prevent the upward drift in weight that can set the stage for metabolic adaptation and weight cycling [12, 17, 19].

Once extra weight has been gained and defended by the brain and body, it becomes much harder to lose and keep off. Instead, small, consistent adjustments over time (rather than drastic diets) are the surest way to maintain a healthy weight and metabolic health.

> **Key Point:** It is much easier to maintain a healthy weight than to lose weight and keep it off after gaining it. Adjust your habits to life changes before extra weight accumulates, and avoid cycles of dieting and regain [17, 19].

1.6.5   Main Takeaway Points

o **Do not assume that weight gain in midlife is due to a slowing metabolism.** For most adults, metabolism remains stable until the 60s [12].

o **Do recognize the role of lifestyle changes.** Weight gain during adulthood is driven primarily by reduced activity and increased calorie intake, not by age-related metabolic decline [17,19].

o **Do understand that weight loss triggers metabolic adaptation.** The body responds to weight loss by slowing metabolism and increasing hunger, making weight regain more likely [17,19].

o **Do avoid repeated cycles of dieting and weight regain.** Weight cycling can further slow metabolism and disrupt metabolic health [17–19].

o **Do prioritize prevention.** Adjust eating and activity habits as life circumstances change to prevent gradual weight gain in the first place [17, 19].

> **Key Point:** Lifelong weight management is easiest when you prevent weight gain before it happens. Once gained, extra weight is defended by powerful biological mechanisms, and repeated cycles of loss and regain can make future weight loss progressively harder [17–19].

1.7   The Influence of Sleep on Weight Gain

The relationship between sleep and body weight has emerged as a significant area of research over the past two decades. While the association between short sleep duration and obesity was initially met with skepticism, a substantial body of evidence now supports sleep as an independent risk factor for weight gain. This section examines the epidemiological evidence, experimental findings, and underlying mechanisms that link insufficient sleep to increased body weight and metabolic dysfunction.

1.7.1   Epidemiological Evidence

Large-scale population studies consistently demonstrate an association between short sleep duration and increased risk of obesity. In a comprehensive systematic review examining 36 studies, Patel and Hu found that short sleep duration appears independently associated with weight gain, particularly in younger age groups [20]. Among pediatric populations, all cross-sectional studies identified showed a positive association between short sleep duration and increased obesity, with remarkably consistent findings across different continents and ethnic groups.

The largest pediatric cohort to date, a Japanese birth cohort of 8,274 children, found that compared to children sleeping 10 hours or more, the odds ratios for obesity were 1.49, 1.89, and 2.89 for sleep durations of 9-10, 8-9, and fewer than 8 hours, respectively [20]. Similar dose-response relationships have been observed in adult populations, though the findings are somewhat more mixed, with 17 of 23 adult cross-sectional studies supporting an independent association between short sleep duration and increased weight.

> **Key Point:** Epidemiological studies consistently show that individuals who sleep less than 6-7 hours per night have increased risk of obesity, with the strongest and most consistent evidence found in children and younger adults [20].

1.7.2 Hormonal Mechanisms

Two key hormones play central roles in appetite regulation and energy balance: leptin and ghrelin. Leptin, produced primarily by fat tissue, acts as a satiety signal, informing the brain when energy stores are sufficient and suppressing appetite. In contrast, ghrelin, produced mainly in the stomach, stimulates hunger and promotes food intake. Under normal circumstances, these hormones work in opposition to maintain energy homeostasis, i.e., leptin levels rise after meals and with increased fat stores, while ghrelin levels rise before meals and during periods of energy deficit. In obesity, this system often becomes dysregulated, with leptin resistance developing despite high leptin levels, and ghrelin responses becoming blunted.

Experimental studies have identified specific hormonal pathways through which sleep deprivation may promote weight gain. The Wisconsin Sleep Cohort Study, involving 1,024 participants, found that short sleep duration was associated with reduced leptin levels and elevated ghrelin levels, independent of body mass index [21]. Participants with habitual sleep durations of 5 hours had 15.5% lower leptin levels compared to those sleeping 8 hours, while those with polysomnographic sleep durations of 5 hours had 14.9% higher ghrelin levels than 8-hour sleepers.

These hormonal changes have direct implications for appetite regulation. Leptin, produced by fat tissue, signals satiety to the brain and suppresses appetite. Ghrelin, primarily produced in the stomach, stimulates hunger and food intake. The combination of reduced leptin and elevated ghrelin with sleep restriction creates a powerful biological drive toward increased caloric intake [21].

Controlled experimental studies have confirmed these associations. Spiegel and colleagues demonstrated that just two days of sleep restriction (4 hours vs. 10 hours) resulted in an 18% decrease in leptin and a 28% increase in ghrelin, accompanied by increased hunger ratings and appetite, particularly for calorie-dense foods with high carbohydrate content [22].

> **Key Point:** Sleep deprivation disrupts key appetite-regulating hormones, with decreased leptin and increased ghrelin creating a biological drive toward overeating that persists even after accounting for body weight [21, 22].

1.7.3   Metabolic Consequences

Beyond appetite regulation, sleep loss has profound effects on glucose metabolism and insulin sensitivity. Tasali and colleagues reviewed evidence showing that both acute total sleep deprivation and chronic partial sleep restriction can induce glucose intolerance and insulin resistance [23]. These metabolic changes occur rapidly, with some studies demonstrating impaired glucose tolerance after just a single night of sleep loss.

The epidemiological evidence linking sleep loss to diabetes is substantial. Several large population studies have found that sleep durations of 6 hours or less, or 9 hours or more, are associated with increased prevalence of diabetes and impaired glucose tolerance, even after adjusting for known diabetes risk factors [23]. Cross-sectional studies using validated sleep quality measures have consistently found associations between poor sleep quality and markers of glucose dysregulation.

The mechanisms underlying these effects involve multiple pathways. Sleep restriction increases evening cortisol levels, enhances sympathetic nervous system activity, and may directly affect pancreatic beta-cell function. These changes promote insulin resistance and impair the body's ability to regulate blood glucose, creating conditions that favor fat storage and weight gain [23].

> **Key Point:** Sleep loss impairs glucose metabolism and insulin sensitivity through multiple pathways, creating metabolic conditions that promote fat storage and increase diabetes risk, independent of changes in body weight [23].

1.7.4 Behavioral and Caloric Intake Effects

Recent controlled feeding studies have provided direct evidence that sleep restriction increases caloric intake. Nedeltcheva and colleagues studied healthy adults in a controlled laboratory environment with ad libitum access to food during periods of 5.5-hour versus 8.5-hour sleep opportunities [24]. Sleep restriction was accompanied by increased consumption of calories from snacks (1,087 vs. 866 kcal/day), with higher carbohydrate content, particularly during nighttime hours from 7 PM to 7 AM.

Importantly, this increased caloric intake was not offset by increased energy expenditure. Total energy expenditure remained similar between sleep conditions (2,526 vs. 2,390 kcal/day), indicating that the additional caloric intake from sleep restriction represented a net positive energy balance [24]. The study found no significant differences in serum leptin and ghrelin levels during the feeding protocol, suggesting that the appetite-stimulating effects of sleep loss may be mediated through mechanisms beyond these classical hormones when food is freely available.

The timing of increased intake is particularly notable, with the greatest increases occurring during extended evening and nighttime hours when sleep-restricted individuals remained awake. This suggests that longer exposure to food availability, combined with altered appetite regulation, contributes to the obesity-promoting effects of insufficient sleep [24].

> **Key Point:** Sleep restriction increases caloric intake, particularly from snacks during extended evening hours, without proportional increases in energy expenditure, leading to positive energy balance and potential weight gain [24].

1.7.5 Clinical Implications

The relationship between sleep and obesity is bidirectional, creating potential for vicious cycles. Obesity increases the risk of sleep disorders, particularly obstructive sleep apnea, which further fragments sleep and may worsen metabolic dysfunction. Studies have shown that sleep apnea is highly prevalent in patients with type 2 diabetes, and conversely, disorders of glucose metabolism are more common in patients with sleep apnea [23].

The clinical implications are profound. The systematic review by Patel and Hu noted that 40% of American adults report obtaining less than 7 hours of sleep, a dramatic increase from historical norms [20]. If chronic partial sleep deprivation contributes causally to obesity, this represents a significant and modifiable public health risk factor.

However, the authors also noted important limitations in the current evidence. Most studies rely on self-reported sleep duration, which may be inaccurate. Only two studies used objective measures (actigraphy), and the possibility of reverse causation, i.e., where obesity causes sleep problems rather than vice versa, cannot be definitively ruled out from observational studies alone [20].

> **Key Point:** Sleep and obesity have a bidirectional relationship, with obesity increasing sleep disorder risk and sleep loss promoting weight gain, creating potential for self-perpetuating cycles of metabolic dysfunction [20, 23].

1.7.6   Main Takeaway Points

- **Do recognize sleep as an independent risk factor for weight gain.** Multiple large-scale studies show consistent associations between short sleep duration and obesity risk, particularly in children and younger adults [20].

- **Do understand the hormonal mechanisms.** Sleep deprivation decreases leptin and increases ghrelin, creating biological drives toward increased food intake that persist even after accounting for body weight [21, 22].

- **Do appreciate the metabolic consequences.** Sleep loss impairs glucose metabolism and insulin sensitivity, promoting conditions that favor fat storage independent of caloric intake [23].

- **Do consider the behavioral effects.** Sleep restriction increases caloric intake, particularly from snacks during extended evening hours, without proportional increases in energy expenditure [24].

   o **Do address sleep disorders in obesity treatment.** The bidirectional relationship between sleep and weight suggests that improving sleep quality may be an important component of weight management strategies [23].

   o **Do not overlook sleep in public health strategies.** With 40% of adults getting insufficient sleep, addressing sleep duration may be a significant opportunity for obesity prevention [20].

> **Key Point:** Mounting evidence from epidemiological, experimental, and mechanistic studies establishes insufficient sleep as a significant risk factor for weight gain through disrupted appetite regulation, impaired metabolism, and increased caloric intake. However, most evidence remains observational, and intervention studies are needed to establish definititive causal relationships and optimal treatment approaches [20–24].

1.8   The Role of Stress in Weight Gain

The relationship between psychological stress and body weight represents one of the most complex and clinically relevant aspects of weight regulation. While the effects of extreme stress on eating behavior have long been recognized, recent research has revealed that even moderate, chronic stress can profoundly influence appetite, food choices, and long-term weight trajectories through multiple interconnected pathways. Understanding these mechanisms is vital for addressing the obesity epidemic, as modern life exposes individuals to unprecedented levels of chronic stress from work, relationships, financial pressures, and social demands.

1.8.1   The Pathways Linking Stress to Weight Gain

Stress activates the hypothalamic-pituitary-adrenal (HPA) axis, leading to the release of cortisol and other stress hormones that directly influence appetite regulation and metabolism. In a prospective study of 339 healthy adults followed for six months, researchers found that higher baseline cortisol levels, insulin levels, and chronic stress each independently predicted greater weight gain over time [25]. Remarkably, nearly half (49.9%) of participants gained weight during this relatively short follow-up period, with those experiencing higher stress showing significantly greater weight increases.

The mechanisms underlying these effects involve complex interactions between stress hormones and appetite-regulating systems. Higher baseline levels of ghrelin, the hormone that stimulates hunger, predicted stronger food cravings six months later, suggesting that stress-induced changes in appetite hormones can have lasting effects on eating behavior [25]. This finding is particularly important because it demonstrates that the relationship between stress and weight is not merely about immediate stress eating, but involves fundamental alterations in the biological systems that control appetite over time.

> **Key Point:** Chronic stress creates lasting changes in appetite-regulating hormones, with higher cortisol, insulin, and ghrelin levels predicting both increased food cravings and weight gain months later [25].

1.8.2 Long-Term Stress and Hedonic Eating

Recent advances in measuring long-term stress exposure through hair cortisol and cortisone levels have provided new insights into how chronic stress influences eating behavior. Hedonic eating refers to eating driven by the pleasure and reward value of food, rather than by physiological hunger or energy needs. This type of eating is characterized by seeking out highly palatable, calorie-dense foods for their rewarding properties, often leading to overconsumption beyond what the body requires for energy balance.

In a study of 108 adults with obesity, researchers found that higher hair cortisone levels (reflecting stress exposure over the previous three months) were significantly associated with stronger food cravings, even after controlling for psychological stress levels [26]. Notably, this association was specific to cortisone rather than cortisol, suggesting that different stress biomarkers may have distinct relationships with eating behavior.

The study also revealed that psychological and biological stress measures can have additive effects on eating behavior. Individuals with both high psychological stress and high biological stress (measured by hair cortisone) showed the highest levels of food cravings and emotional eating, indicating that these different aspects of stress may compound each other's effects on eating behavior [26]. This finding has important implications for understanding why some individuals are more susceptible to stress-related weight gain than others.

Key Point: Long-term biological stress, measured through hair cortisone levels, independently predicts stronger food cravings in people with obesity, with psychological and biological stress showing potentially additive effects on hedonic eating behaviors [26].

1.8.3   Individual Differences in Stress Responsivity

Not all individuals respond to stress in the same way, and these differences have important implications for weight regulation. Research comparing emotional eaters to non-emotional eaters under controlled stress conditions revealed striking differences in both physiological and neural responses [27]. When exposed to an acute laboratory stressor, emotional eaters showed significantly greater increases in both cortisol and anxiety levels compared to non-emotional eaters, indicating a hyperactive stress response system.

Perhaps even more revealing were the differences in brain activity during food reward processing. Following stress exposure, emotional eaters showed significantly reduced activation in key reward regions of the brain, including the caudate, nucleus accumbens, and putamen, when anticipating food rewards [27]. This pattern suggests that stress may impair the normal reward processing that helps regulate food intake, potentially leading to compensatory overeating as individuals attempt to achieve the same level of reward satisfaction.

Interestingly, despite these dramatic differences in stress reactivity and brain function, the two groups did not differ in actual food consumption during the laboratory session. This finding highlights the complexity of the stress-eating relationship and suggests that the effects of stress on eating behavior may be most apparent in real-world settings where food choices and eating patterns unfold over longer time periods [27].

Key Point: Individuals prone to emotional eating show both hyperactive stress responses (elevated cortisol and anxiety) and impaired brain reward processing under stress, creating biological vulnerability to stress-related eating problems even when immediate food intake appears normal [27].

1.8.4   The Stress-Sleep-Eating Cycle

One of the most important pathways through which stress influences weight regulation is through its effects on sleep, which in turn affects eating behavior. A comprehensive study of 317 university students during exam periods revealed strong positive correlations between perceived stress levels and insomnia symptoms, with the strongest associations found between feeling out of control in life and frequent nighttime awakenings [28].

The same study demonstrated equally strong relationships between stress levels and disordered eating behaviors, including overeating, lack of control over eating, binge eating episodes, and unhealthy food choices. Crucially, the research identified insomnia as a mediator in the relationship between stress and disordered eating [28]. When insomnia was included in statistical models, the direct effect of stress on eating behavior became non-significant, while insomnia remained a strong predictor of eating problems. This suggests that much of stress's impact on eating behavior may be mediated through its disruptive effects on sleep.

Students experiencing high levels of stress reported significantly greater difficulties with sleep initiation and higher levels of disordered eating compared to their lower-stress peers. The magnitude of these differences was substantial, with highly significant statistical differences in both insomnia scores and eating behavior scores between high- and low-stress groups [28].

> **Key Point:** Insomnia acts as a vital mediator between stress and disordered eating, suggesting that stress-related sleep disruption may be a primary pathway through which stress leads to weight gain and eating problems [28].

1.8.5   Implications for Weight Management

The research on stress and weight regulation has profound implications for both prevention and treatment of obesity. The finding that stress effects on eating behavior are mediated through sleep disruption suggests that interventions targeting sleep quality may be particularly effective for preventing stress-related weight gain. Similarly, the identification of individual differences in stress responsivity indicates that personalized approaches may be necessary, with some individuals requiring more intensive stress management support than others.

The discovery that biological and psychological stress can have additive effects suggests that comprehensive assessment of stress exposure—including both subjective reports and objective biomarkers—may be valuable for identifying individuals at highest risk for stress-related weight gain [26]. For those already struggling with obesity, addressing chronic stress through evidence-based interventions may be an important component of weight management programs. The research also highlights the importance of addressing the underlying systems that regulate appetite and reward processing, rather than focusing solely on conscious food choices. Since stress appears to alter fundamental brain reward circuits and appetite-regulating hormones, interventions that target these biological systems (such as stress reduction, sleep optimization, and mindfulness-based approaches) may be more effective than traditional dietary counseling alone [27].

> **Key Point:** Effective weight management must address stress as a fundamental driver of weight gain through its effects on sleep, appetite hormones, and brain reward systems, rather than treating eating behavior as merely a matter of conscious choice [25–28].

1.8.6 Main Takeaway Points

○ **Do recognize stress as a major biological driver of weight gain.** Chronic stress alters appetite-regulating hormones and predicts weight gain months later, independent of conscious eating behaviors [25].

○ **Do understand that stress effects vary dramatically between individuals.** Some people show hyperactive stress responses and impaired brain reward processing that make them particularly vulnerable to stress-related eating problems [27].

○ **Do address sleep as a critical link between stress and eating.** Insomnia mediates much of the relationship between stress and disordered eating, making sleep optimization a key intervention target [28].

○ **Do consider both psychological and biological measures of stress.** Long-term biological stress markers like hair cortisone can predict eating problems independently of reported psychological stress [26].

○ **Do not underestimate the long-term effects of moderate stress.** Even routine life stress can create lasting changes in appetite regulation that promote weight gain over time [25, 26].

○ **Do not assume that immediate eating behavior reflects underlying vulnerability.** Stress-related changes in brain reward processing may create risk for eating problems that don't manifest immediately in laboratory settings [27].

> **Key Point:** Stress influences weight regulation through multiple interconnected pathways involving appetite hormones, sleep disruption, and altered brain reward processing. Understanding these mechanisms reveals why traditional approaches focusing solely on conscious food choices often fail, and why comprehensive stress management is essential for long-term weight regulation [25–28].

1.9 How Body Ideals Shape Our Diets

The influence of body ideals on eating behaviors and diet quality has been firmly established by a growing body of international research. These ideals (shaped by societal, cultural, and media-driven narratives) do not merely reside in the realm of self-image, but actively shape what, how, and why we eat. The following section synthesizes recent evidence demonstrating the complex interplay between body image, social media, and nutritional behaviors in diverse adult populations.

1.9.1 Body Image and Eating Behaviors

A consistent finding across cross-sectional and longitudinal studies is that negative body image is closely linked to maladaptive eating patterns in adults of all ages. Large-scale survey of 3,100 adults, for example, found that adults with lower body image satisfaction were more likely to engage in uncontrolled eating, emotional eating, and heightened susceptibility to hunger, while those with higher body satisfaction demonstrated greater cognitive restraint (i.e., conscious dietary control); suggesting a more conscious, regulated approach to eating [29]. These relationships remained significant even after controlling for age and gender, underscoring the robust association between negative body image and disordered or unhealthy eating behaviors.

Similarly, work in Iran and New Zealand confirms that positive body image and body esteem are associated with healthier eating attitudes, while low body esteem and negative body image predict unhealthy eating patterns and greater risk for eating disorders [30, 31]. These findings are echoed in multi-country samples and across cultural settings, indicating that the impact of body ideals on eating behaviors is a global phenomenon.

> **Key Point:** Negative body image is a reliable predictor of emotional eating, uncontrolled eating, and maladaptive nutritional behaviors, while positive body image and body esteem are associated with adaptive eating attitudes and dietary restraint [29, 30].

1.9.2 The Role of Social Media

Modern body ideals are increasingly shaped by social media, which amplifies and normalizes narrow standards of beauty and fitness. Multiple systematic reviews and experimental studies demonstrate that exposure to idealized images on platforms like Instagram, TikTok, and Facebook increases body dissatisfaction and the internalization of unrealistic appearance standards, which in turn drive restrictive dieting, emotional eating, and food guilt [32–34]. These effects are mediated by upward social comparison and the pressure to conform to peer and influencer-promoted body ideals.

Crucially, the impact of social media is not limited to women; although women are more likely to report negative effects, men also internalize muscularity ideals, which is reflected in their food choices and attitudes toward eating [35]. Qualitative studies reveal that young adults experience both direct and indirect pressure to modify their diets and appearance, and that social media often fosters confusion and anxiety regarding what constitutes "healthy" eating [31, 32].

> **Key Point:** Social media acts as a powerful vector for body ideals, intensifying body dissatisfaction and promoting restrictive or disordered eating behaviors through mechanisms of social comparison and internalization [32–34].

1.9.3 Cultural and Family Influences

While media and peers are well-known drivers of body image and eating behaviors, research from non-Western contexts highlights that family attitudes and cultural values are equally—sometimes even more—important in shaping body perceptions and eating habits. Recent studies from Saudi Arabia and Iran, using large samples and validated questionnaires, explored these relationships in depth. [30, 35].

In Saudi Arabia, Alburkani and colleagues surveyed over 600 university students and found that 41% were at risk for eating disorders, with a quarter having significant body shape concerns. Notably, family attitudes toward appearance were just as influential as social media exposure. Students whose families emphasized appearance, weight, and thinness (such as making frequent comments about weight or dieting) were more likely to feel dissatisfied with their bodies and have disordered eating attitudes. These patterns appeared in both men and women, though women reported higher rates of problematic eating and more social media use. The study also noted that in rapidly Westernizing societies, traditional values interact with global beauty standards, sometimes intensifying pressures to conform to unrealistic ideals.

Similarly, Sharif-Nia and colleagues in Iran surveyed over 750 adolescents and adults. They found that higher body esteem and positive body image strongly predicted healthy eating attitudes, while negative body image was linked to unhealthy eating patterns, including restrictive dieting. Demographic factors like age or income had little effect; instead, body esteem and positive body image were the strongest predictors. The study also emphasized the cultural significance of food in Iran and how pressure to meet beauty standards—often magnified by media—can lead to body dissatisfaction and unhealthy eating behaviors.

These studies show that family and cultural influences are central to understanding body image and eating behaviors. Parental emphasis on appearance and frequent comments about weight can increase the risk of body dissatisfaction and disordered eating, while positive family environments and cultural messages that value body diversity and health can be protective. Both studies highlight the need for interventions that address family dynamics and promote body positivity within the cultural context, rather than focusing only on media or peer influences.

> **Key Point:** Family and cultural context have a strong influence on body image and eating behaviors. Parental emphasis on thinness, frequent appearance-related comments, and broader societal beauty standards can increase the risk of body dissatisfaction and disordered eating, especially in rapidly changing societies.

1.9.4 The Potential of Body-Positive Content

Emerging research on body-positive social media content suggests that exposure to diverse, body-affirming images can improve mood, body satisfaction, and body appreciation in the short term, and may buffer against some of the adverse impacts of conventional beauty standards.

A recent comprehensive review led by Jiménez-García and colleagues (2025) looked at dozens of studies from around the world to find out whether body-positive content on social media actually helps people feel better about their bodies [34]. The researchers analyzed findings from 56 studies, including both short-term experiments and longer-term investigations, exploring how exposure to body-positive images, videos, and messages affects people's sense of body satisfaction and emotional well-being.

The evidence shows that viewing body-positive content (especially posts that celebrate body diversity and encourage self-acceptance) can give an immediate boost to how people feel about their bodies. People who saw these kinds of posts reported feeling more satisfied and happier with their appearance, and their moods also improved. These positive effects were seen not just in brief experiments, but also in studies where people engaged with body-positive content regularly over weeks. The more often people saw diverse and accepting images, the more likely they were to develop a lasting appreciation for their own bodies.

However, the research also found that the benefits of body-positive content have their limits. While such content can help people feel better in the moment, it doesn't always reduce deeper issues, such as the tendency to constantly compare oneself to others or to be overly focused on appearance. For some, the focus on appearance (even in a positive light) can still leave them stuck in a cycle of self-scrutiny or comparison. In other words, while body-positive posts may lift spirits and encourage acceptance, they are not a complete cure for the pressures and anxieties created by traditional beauty standards.

Interestingly, the effectiveness of body-positive content depends on how it is presented. Posts that combine images of diverse bodies with supportive, self-compassionate captions tend to be more effective than images or captions alone. On the other hand, if body-positive posts still emphasize appearance or subtly reinforce mainstream beauty ideals, they can sometimes miss the mark or even backfire.

Most of the research in this area has focused on young women, who are often the most affected by social media beauty pressures. But the available evidence suggests that body-positive content can be helpful for people of all genders and ages, though more research is needed to understand its impact across different groups. As the body-positive movement continues to grow, it will be important for content creators, health professionals, and educators to focus on messages that encourage self-acceptance, compassion, and a broader view of beauty.

> **Key Point:** Body-positive social media content can improve body satisfaction and mood, but alone may not be sufficient to dismantle deeper patterns of comparison and self-objectification [34].

1.9.5 Complex, Bidirectional Relationships

Collectively, these findings point to a dynamic, bidirectional relationship between body ideals and eating behaviors. On one hand, the internalization of narrow and unrealistic beauty standards (often perpetuated by conventional and social media) triggers disordered eating patterns, while on the other, maladaptive eating behaviors and chronic dieting can further erode body satisfaction and perpetuate negative self-image [29, 30, 33]. This cycle is reinforced by media, peer, and family influences, and may be buffered (but not eliminated) by positive body image interventions and critical media literacy.

It is especially important for young people to recognize that the body ideals promoted by social media and advertising are frequently artificial and unattainable. Many of the images that saturate platforms like Instagram, TikTok, and fitness or fashion magazines depict models in extreme, unsustainable states: their photos are often digitally retouched to erase natural imperfections, artificially accentuate muscle definition, and shrink waistlines. In some cases, models follow dehydration protocols before photo shoots to exaggerate muscle visibility, particularly abs, making their appearance even more unrealistic and unhealthy [31, 32].

Such images do not represent the day-to-day reality of healthy bodies. Instead, they are curated, filtered, and manipulated to conform to a fleeting and commercialized aesthetic ideal. When young people internalize these standards (believing that retouched and dehydrated bodies are the norm or the goal) they may develop body dissatisfaction, emotional distress, and harmful eating behaviors such as extreme dieting, restriction, bingeing, or emotional eating [30, 32, 33]. Studies confirm that exposure to these unrealistic standards, especially through social media, increases comparison and the pressure to diet or "fix" one's body, leading to a cycle of striving for an unattainable ideal and recurring disappointment [32, 33, 35].

This process is particularly pronounced among young women, but increasing evidence shows that young men are also affected, pressured to display hyper-muscular or lean physiques that are equally manipulated and staged [32, 35]. The resulting dissatisfaction is not just an individual problem; it is a public health concern, with higher rates of disordered eating and body image disturbance observed in populations most exposed to these ideals [29, 33, 35].

To break this cycle, it is vital for young people to build media literacy skills, i.e., learning to recognize that much of what they see online or in magazines is constructed, edited, and often unhealthy in reality. Developing a positive body image and critical awareness offers protection: individuals who appreciate and accept their bodies as they are, regardless of societal pressures or filtered imagery, are more likely to engage in healthy, balanced eating and avoid the dangers of chronic dieting or disordered eating [30, 34].

> **Key Point:** Young people must be aware that the "perfect" bodies seen on social media and in advertisements are often the result of digital editing, manipulation, and unhealthy practices like dehydration. These are not standards to admire or pursue. Instead, cultivating media literacy and a positive, realistic appreciation of one's own body is vital to resisting harmful pressures and fostering healthy eating behaviors [30, 32, 33, 35].

1.9.6   Main Takeaway Points

○ **Do recognize that body ideals, especially those amplified by social media, shape not just self-image but also everyday eating behaviors.**

○ **Do understand that negative body image is a risk factor** for maladaptive, emotional, and uncontrolled eating, while positive body image is protective.

○ **Do consider the role of social comparison and internalization in mediating the impact of media and peer influences on diet.**

○ **Do not assume all body-positive content is equally effective**; interventions should address underlying drivers of comparison and self-objectification.

○ **Do address family and cultural context in interventions**, as these factors can buffer or exacerbate media-driven pressures.

> **Key Point:** Body ideals (shaped by media, peers, and family) directly influence eating behaviors, diet quality, and risk for disordered eating. The relationship is complex, bidirectional, and shaped by social comparison, internalization, and cultural context. Effective intervention requires a multifaceted approach.

1.10 Boredom and Weight Gain

While much attention has been paid to the roles of stress, sleep, and body image in weight regulation, a growing body of research highlights the unique and powerful influence of boredom on eating behavior and long-term weight gain. Unlike other negative emotions, boredom is a distinct affective state characterized by the aversive experience of wanting, but being unable, to engage in satisfying activity [36,37]. This section synthesizes recent evidence on how boredom drives eating, the mechanisms involved, and its implications for obesity and mental health.

1.10.1 Defining Boredom

Boredom is not simply a lack of stimulation or a mild form of apathy. Empirical research demonstrates that boredom is a unique emotional experience, distinct from apathy, anhedonia, or depression [37]. The core of boredom lies in an attentional failure: the inability to engage with internal or external stimuli in a way that feels meaningful or satisfying [36]. This state is often accompanied by restlessness, dissatisfaction, and a heightened awareness of the gap between one's current experience and desired engagement.

> **Key Point:** Boredom is a discrete, aversive emotional state marked by a lack of engagement and meaning, and is empirically distinct from depression or apathy [36,37].

1.10.2   Boredom as a Driver of Eating Behavior

Multiple studies have established that boredom is a robust predictor of increased food intake, independent of other negative emotions such as sadness or stress [38]. In a week-long diary study, state boredom was found to positively predict daily calorie, fat, carbohydrate, and protein consumption, even after controlling for stress, enjoyment, and trait boredom proneness [38]. Experimental manipulations confirm this causal relationship: participants exposed to boring tasks reported greater desire to snack and consumed more unhealthy foods compared to those in non-boring conditions, especially among individuals high in self-awareness [38].

Importantly, boredom does not simply increase all forms of eating. It tends to promote the consumption of more exciting or stimulating foods (often those that are high in sugar, fat, or novelty) rather than bland or unexciting healthy options [38]. However, when healthy foods are made more exciting (e.g., cherry tomatoes vs. plain crackers), bored individuals may also increase their intake of these alternatives, suggesting that the drive is for stimulation rather than unhealthiness per se.

> **Key Point:** Boredom specifically increases the desire for and consumption of stimulating, often unhealthy foods, as a means of escaping the aversive state of disengagement [38].

Not everyone is equally susceptible to boredom-induced eating. Individuals with high boredom proneness and difficulties in emotion regulation are at greater risk for emotional and inappropriate eating, including eating in response to boredom, negative emotions, and external cues [39]. These traits may help identify those most vulnerable to weight gain and unhealthy eating patterns in environments where boredom is common.

> **Key Point:** Boredom proneness and poor emotion regulation are significant risk factors for emotional eating and may contribute to unhealthy weight gain [39].

1.10.3   Mechanisms: Escaping the Bored Self

The motivation to eat when bored appears to be less about seeking pleasure and more about escaping the discomfort of self-awareness and meaninglessness [38]. Boredom signals a lack of purpose, prompting individuals to seek activities (such as eating) that distract from this aversive self-focus.

Experimental evidence shows that, under boredom, people are not only more likely to eat, but also more willing to engage in other sensation-seeking behaviors, even those that are aversive (such as self-administering mild electric shocks) [40]. This supports the idea that the primary drive is to disrupt monotony and escape the unpleasant state of boredom, rather than to obtain positive reward.

> **Key Point:** Eating in response to boredom is primarily motivated by the desire to escape aversive self-awareness and monotony, not simply to seek pleasure [38, 40].

The COVID-19 pandemic and associated lockdowns provided a natural experiment in the effects of boredom and disrupted routines on eating behavior. In a large Italian survey, many individuals reported increased consumption of comfort foods, sweets, and snacks during lockdown, alongside reduced physical activity and perceived weight gain [41]. While some improved their diet quality by eating more fruits and vegetables, a significant proportion turned to food as a coping mechanism for boredom and emotional distress.

> **Key Point:** Periods of enforced inactivity and boredom, such as during pandemic lockdowns, are associated with increased emotional and boredom-driven eating, contributing to weight gain and unhealthy dietary patterns [41].

1.10.4 Main Takeaway Points

- o **Do recognize boredom as a distinct and powerful driver of eating behavior.** Boredom increases the desire for stimulating foods and promotes eating as a means of escaping aversive self-awareness [38, 40].

- o **Do understand that boredom-induced eating is not simply about pleasure.** The primary motivation is to disrupt monotony and avoid the discomfort of disengagement [40].

- o **Do identify individuals at risk.** Those high in boredom proneness and with poor emotion regulation are especially vulnerable to emotional and boredom-driven eating [39].

- o **Do consider the broader context.** Societal disruptions and periods of inactivity can amplify boredom and its effects on eating, highlighting the need for adaptive coping strategies [41].

○ **Do integrate interventions.** Addressing boredom, improving emotion regulation, and promoting engaging, meaningful activities may help prevent emotional eating and its contribution to weight gain and depression [42, 43].

> **Key Point:** Boredom is a unique and underappreciated driver of emotional eating and weight gain. Effective prevention and intervention require strategies that address the underlying need for engagement and meaning, not just dietary restraint.

1.11   The Cycle of Emotional Eating

Emotional eating (defined as eating in response to negative emotions such as stress, sadness, or boredom rather than physiological hunger) has emerged as a central behavioral mechanism linking mood disorders to long-term weight gain. Recent large-scale, prospective, and genetic studies provide robust evidence that this cycle is not only common but also bidirectional and self-reinforcing, with important implications for both weight management and mental health.

1.11.1   The Bridge Between Depression and Weight

A landmark 7-year prospective cohort study from Finland, known as the Dietary, Lifestyle and Genetic determinants of Obesity and Metabolic syndrome (DILGOM) study, directly tested whether emotional eating mediates the relationship between depression and subsequent weight gain [42]. In this study, over 5,000 adults aged 25 to 74 were followed from 2007 to 2014, with depressive symptoms measured using the Center for Epidemiological Studies Depression Scale and emotional eating assessed via the Three-Factor Eating Questionnaire. Body mass index (BMI) and waist circumference were measured at both baseline and follow-up.

The results were striking: both depression and emotional eating independently predicted greater increases in BMI and waist circumference over the seven-year period. More importantly, structural equation modeling revealed that emotional eating significantly mediated the effect of depression on weight gain and abdominal obesity. This mediation effect was especially pronounced in women and younger adults, highlighting the particular vulnerability of these groups.

The study also uncovered a vital moderating role for sleep. Night sleep duration significantly influenced these associations: emotional eating predicted higher weight gain only among those with shorter sleep (seven hours or less per night), but not among longer sleepers (nine hours or more). Physical activity, while associated with lower emotional eating, did not significantly moderate the effect of emotional eating on weight gain. These findings provide strong longitudinal evidence that emotional eating is a key behavioral pathway through which depression leads to weight gain, particularly in individuals with poor sleep [42].

> **Key Point:** Emotional eating mediates the relationship between depression and long-term weight gain, especially in women, younger adults, and those with short sleep duration [42].

1.11.2   Mood and Food: A Two-Way Street

Recent cross-sectional and experimental studies further clarify the bidirectional nature of the relationship between mood and eating. For example, a 2025 study of over 500 university students used validated questionnaires to assess food preferences, depression severity (Beck Depression Inventory), and physical activity [44]. The study found that students with higher depression scores had significantly lower preferences for nutrient-dense foods such as grains, fruits, and vegetables, and higher preferences for snacks and energy-dense foods. Logistic regression analysis showed that higher preferences for fruits and vegetables were associated with reduced depression risk, while a preference for snacks increased risk, even after adjusting for age, gender, BMI, smoking, and activity.

Notably, the relationship was bidirectional: not only did depression predict unhealthy food preferences, but food preferences also predicted depression risk. This highlights a feedback loop in which emotional states and food choices reinforce each other, perpetuating both poor diet and poor mental health [44].

> **Key Point:** Depression is associated with lower preference for healthy foods and higher preference for snacks; conversely, healthy food preferences are linked to lower depression risk, supporting a bidirectional relationship [44].

1.11.3 Diet Composition and Mental Disorders

To move beyond associations and test for causality, a 2025 bidirectional
Mendelian randomization study (i.e., a method that uses genetic variants
to infer whether an observed association is likely to be causal) used genetic
instruments to examine the relationships between macronutrient intake and
mental disorders in large European cohorts [45]. In Mendelian randomization,
genetic variants associated with dietary intake are used as proxies for long-
term exposure, helping to minimize confounding and reverse causation
that often limit traditional observational studies. The results showed that
genetically predicted high protein and high fat intake increased the risk of
anxiety and depression, while high carbohydrate intake was associated with
increased risk of bipolar disorder. In the reverse direction, genetic liability
to neuroticism increased sugar intake, and schizophrenia was associated
with lower intake of all macronutrients. While most associations were
nominally significant, the trends support a causal link between unbalanced
macronutrient intake and psychiatric risk. This genetic evidence strengthens
the case that dietary patterns can contribute to the development of mood
disorders, not just result from them [45].

> **Key Point:** Genetic evidence supports a causal effect of high
> protein and fat intake on anxiety and depression risk, and a
> bidirectional relationship between diet and mental health [45].

The protective role of healthy dietary patterns is further supported by a
comprehensive 2019 meta-analysis and systematic review, which synthesized
data from 41 observational studies (20 longitudinal, 21 cross-sectional) examining the link between healthy dietary indices and depressive outcomes [43].
The review found that highest adherence to a Mediterranean diet was associated with a 33% lower risk of incident depression. Lower scores on the
Dietary Inflammatory Index (i.e., less pro-inflammatory diets) were also
associated with lower depression risk. Other healthy eating indices, such as
the Healthy Eating Index and the Dietary Approaches to Stop Hypertension
(DASH) diet (fruits, vegetables, whole grains, low-fat dairy, and limited
saturated fat, red meat, and sugar) showed similar trends.

The evidence was strongest for Mediterranean and anti-inflammatory
diets, and the association was robust in longitudinal studies. These findings
suggest that healthy dietary patterns may confer protection against depression, while unhealthy, pro-inflammatory, or energy-dense diets may increase
risk [43].

Key Point: Adherence to healthy dietary patterns, especially
the Mediterranean diet, is associated with lower risk of depression, supporting dietary interventions as part of mental health
strategies [43].

1.11.4   Sleep, Stress, and Eating

The studies above, along with mechanistic research, suggest several pathways
for the cycle of emotional eating. Neurotransmitter modulation is one such
pathway: carbohydrates can increase serotonin, improving mood, while
protein and fat intake may affect dopamine and gamma-aminobutyric acid
(GABA), i.e., the primary inhibitory neurotransmitter in the brain, which
helps regulate neuronal excitability and is linked to anxiety and mood
regulation, thereby influencing reward and anxiety [44, 45]. Unhealthy diets
promote inflammation and oxidative stress, both of which are implicated
in depression pathophysiology [43]. The gut-brain axis also plays a role:
dietary fiber and plant foods support a healthy gut microbiome, which
communicates with the brain and influences mood [43, 44]. Finally, sleep
and stress are critical: short sleep duration amplifies the effect of emotional
eating on weight gain, and stress can trigger both poor sleep and emotional
eating [42].

Key Point: Emotional eating is driven by complex interactions
between mood, neurotransmitters, inflammation, sleep, and the
gut-brain axis, creating a self-reinforcing cycle of poor diet and
poor mental health.

1.11.5   Breaking the Cycle

The evidence suggests that breaking the cycle of emotional eating requires
integrated approaches. Interventions that teach emotion regulation, mind-
fulness, and coping skills can reduce emotional eating and its impact on
weight and mood [42]. Promoting Mediterranean or anti-inflammatory diets
may reduce depression risk and improve overall well-being [43]. Improving
sleep duration and managing stress can weaken the link between emotional
eating and weight gain [42]. Individuals with high emotional eating, short
sleep, or high stress may need tailored interventions.

Key Point: Effective prevention and treatment of weight gain and depression should address emotional eating, diet quality, sleep, and stress in an integrated, personalized manner.

1.11.6   Main Takeaway Points

○ **Do recognize emotional eating as a key mechanism linking depression and weight gain.** Longitudinal and genetic studies show that emotional eating mediates the effect of depression on obesity, especially in women and short sleepers [42].

○ **Do understand the bidirectional relationship between mood and food choices.** Depression predicts unhealthy food preferences, and unhealthy food preferences increase depression risk [44].

○ **Do note the causal role of diet composition in mental health.** High protein and fat intake increase risk for anxiety and depression, while healthy dietary patterns are protective [43, 45].

○ **Do address sleep and stress as amplifiers of emotional eating.** Short sleep duration strengthens the link between emotional eating and weight gain [42].

○ **Do integrate emotion regulation, diet quality, and sleep interventions.** Multifaceted, personalized approaches are needed to break the cycle of emotional eating and improve both weight and mental health outcomes.

> **Key Point:** Emotional eating is a central behavioral link between depression and weight gain, reinforced by poor diet, sleep, and stress. Breaking this cycle requires integrated strategies targeting emotion regulation, healthy eating, and sleep hygiene, with special attention to those most at risk.

Chapter 2

The Psychology of Health

2.1 The Housekeepers Study

A central question in health psychology is the extent to which our beliefs and expectations, i.e., our *mindset*, can influence physical health outcomes, independent of actual behavioral change. The landmark study by Langer and Crum (2007), titled "Mind-Set Matters: Exercise and the Placebo Effect," provides compelling evidence that simply changing how individuals perceive their daily activities can lead to measurable improvements in health. This study is frequently cited as a demonstration of the placebo effect in the context of exercise and weight management, and it has had a significant impact on the broader scientific understanding of the mind-body connection [46].

2.1.1 Study Design and Methodology

The study was conducted as a field experiment in seven hotels in the United States, involving 84 female room attendants (housekeepers) aged 18 to 55. The majority of participants were Hispanic or African American. The hotels were randomly assigned to either the intervention or control condition, with four hotels (n=44) in the intervention group and three hotels (n=40) in the control group. This cluster randomization was chosen to prevent cross-contamination between groups, though it introduces potential confounding variables related to hotel-specific factors [46].

At baseline, all participants completed assessments of physiological health variables known to be affected by exercise, including weight, body mass index (BMI), waist-to-hip ratio, and blood pressure. They also reported their perceived amount of exercise and general health.

2.1.2   Intervention

The intervention was purely informational. Housekeepers in the intervention group attended a 15-minute presentation in which researchers explained that their daily work (cleaning hotel rooms) constituted significant physical activity and met or exceeded the Surgeon General's recommendations for an active lifestyle. The presentation included specific examples, such as "working at the gym on this machine is like making a bed," to help participants reframe their work as exercise. Informational posters and letters reinforcing this message were also placed in staff areas. In contrast, the control group received no such information and continued their work as usual [46].

2.1.3   Results

Four weeks after the intervention, both groups were reassessed using the same physiological and self-report measures. Notably, there were no reported changes in actual behavior: neither group altered their eating habits or work routines, and no increase in physical activity outside of work was observed. However, the intervention group experienced a significant shift in mindset, i.e., they now perceived themselves as getting more exercise than before.

This change in perception was accompanied by measurable improvements in health outcomes for the intervention group compared to the controls, namely reductions in weight, body fat, waist-to-hip ratio, BMI, and blood pressure. These results were achieved without any actual increase in physical activity, supporting the hypothesis that the health benefits of exercise may be mediated, in part or in whole, by the placebo effect; that is, by the belief that one is engaging in health-promoting behavior [46].

2.1.4   Interpretation and Implications

As Langer herself explained in subsequent interviews, "We have two groups. One group that now knows their work is exercise. The other group that doesn't realize their work is exercise. We take hosts of measures before we start. And at the end, the two groups are not eating any differently. One group isn't working any harder. They're basically the same, except one group believes their work is exercise. As a result of that change in mind, they lost weight. There was a change in waist to hip ratio, body mass index, and their blood pressure came down."

This study provides striking evidence that our minds and bodies are deeply interconnected. If our mindset is tuned to see our daily activities as beneficial for our health, this belief alone can produce real, measurable improvements in physiological health markers. The findings have broad implications for public health messaging, workplace wellness programs, and the design of interventions aimed at improving health outcomes through psychological as well as behavioral means.

2.1.5   Scientific Context and Critique

The Langer and Crum study is a cornerstone in the literature on the placebo effect and the power of mindset in health. It has inspired further research into how beliefs about exercise, diet, and stress can shape physical outcomes. However, it is important to note that subsequent attempts to replicate the findings have yielded mixed results, and methodological critiques (such as the use of cluster randomization and the short duration of follow-up) have been raised. Nonetheless, the study remains a powerful illustration of the potential for mindset interventions to complement traditional approaches to dieting, fitness, and health.

> **Key Point:** Believing that your everyday activities (such as walking, taking the stairs, gardening, or commuting) count as exercise can have a positive impact on your overall health and well-being. This mindset shift applies to all kinds of daily routines, not just your job or household chores, and can influence motivation, physical outcomes, and even how you perceive your own fitness, regardless of whether your actual activity levels change. In other words, viewing regular movements throughout your day as exercise, rather than dismissing them as insignificant, can lead to real psychological and physiological benefits.

2.2   The Perceived Time and Blood Sugar

A notable demonstration of the mind-body connection comes from research investigating whether our *perception* of time (rather than actual time elapsed) can influence physiological processes such as blood glucose regulation. The study by Park, Pagnini, Reece, Phillips, and Langer (2016), titled "Blood sugar level follows perceived time rather than actual time in people with type 2 diabetes," provides some evidence that psychological expectations surrounding the passage of time can shape core metabolic outcomes [47].

2.2.1   Study Design and Methodology

This experiment recruited 46 adults (mean age 54.1 years, 24 women) with type 2 diabetes of at least 12 months' duration, all managed by diet and/or metformin. To minimize variability in blood glucose levels (BGLs), participants were instructed to fast for at least 8 hours (mean fasting time was nearly 12 hours) before the laboratory session. All participants completed a weeklong pre-study diary of their glucose levels for familiarization.

Upon arrival, participants surrendered all devices that could display the time. They were told that the experiment was about cognitive functioning in diabetes, masking the true purpose of the study. Participants were then randomly assigned to one of three groups:

○ **Normal Time Condition:** The desk clock displayed actual time as 90 minutes passed.

○ **Slow Time Condition:** The clock was rigged to run at half speed, so only 45 minutes appeared to pass over the actual 90 minutes.

○ **Fast Time Condition:** The clock ran at double speed, so it showed 180 minutes passing over the actual 90 minutes.

During the study, participants played simple video games for the entire period, with prompts to switch games every 15 minutes (as indicated by their respective clocks), ensuring they attended to the passage of time. All groups had their blood glucose measured immediately before and after the task period. At the end, participants were asked to estimate how much time had passed (manipulation check), and to rate their current stress and hunger levels.

2.2.2   Results

The time manipulation was highly effective: participants in the Slow, Normal, and Fast clock conditions estimated that 59, 112, and 175 minutes had passed, respectively, closely matching the intended clock manipulations.

Crucially, *changes in blood glucose levels tracked participants' perceptions of time, not actual elapsed time.* Those who believed more time had passed (Fast condition) had a greater decrease in BGL (mean: 23.5 mg/dL) compared to the Normal group (15.1 mg/dL), while those who believed less time had passed (Slow condition) had a smaller decrease (9.8 mg/dL). Across all participants, the amount of BGL decrease was significantly correlated with perceived elapsed time (correlation $r = 0.53$, $p < 0.01$).

Importantly, self-reported stress did not differ between groups, and standard hunger ratings only increased in the Fast group (consistent with believing more time had passed since their last meal). The differences in BGL were not attributable to activity levels or other confounding variables, as all groups performed similar, low-exertion tasks throughout.

2.2.3 Interpretation and Implications

This study provides clear evidence that the body's metabolic rhythms (in this case, blood glucose regulation in people with diabetes) can be shaped by abstract psychological constructs such as the perception of time. These findings challenge the classical biomedical assumption that glucose levels depend solely on biological processes and external behaviors. Instead, they suggest that cognitive expectations (how much time we believe has passed) can modulate physiological responses [47].

For diabetes management, this opens up new possibilities: addressing patients' expectations and psychological framing of time may help optimize metabolic control, alongside traditional dietary and pharmacological interventions. More broadly, it underscores the need for an integrated biopsychosocial approach to chronic disease.

2.2.4 Scientific Context and Critique

The Perceived Time and Blood Sugar Study builds on prior research showing that expectation and mindset can alter physiological markers, such as the classic placebo effect or changes in satiety hormones. It stands out by demonstrating that even something as fundamental as the internal clock can recalibrate the body's chemistry. While the manipulation was effective and confounders were well controlled, generalizability to non-diabetic or more diverse populations remains to be established. Future research should explore how these findings translate to everyday life and long-term health outcomes.

Key Point: Blood glucose levels after fasting are influenced more by *how much time they think has passed* than by the actual time elapsed. This highlights the profound role of mindset and expectation: even your body's metabolic rhythms can follow your beliefs about time.

2.3 The Perceived Sugar Study

While the previous study demonstrated that our perception of time can directly influence blood glucose regulation [47], related research has shown that our beliefs about what we consume (especifically, the perceived sugar content of food) can similarly shape metabolic responses, independent of the food's actual composition.

A growing body of research in health psychology suggests that our beliefs and expectations (i.e., our *mindset*) can influence not only subjective experiences but also measurable physiological responses. While the Housekeepers Study demonstrated that perceiving daily activity as "exercise" can lead to improved health markers, a recent investigation by Park, Pagnini, and Langer (2020) extended this paradigm into the realm of glucose metabolism in people with type 2 diabetes. Their study, titled "Glucose metabolism responds to perceived sugar intake more than actual sugar intake," provides striking evidence that what people *believe* about the sugar content of their food can have a more pronounced effect on their blood glucose than the food's actual sugar content [48].

2.3.1 Study Design and Methodology

The study was a laboratory-based, within-subject experiment conducted at Harvard University. Thirty adults (mean age 52.1 years, 47% women) with insulin-independent type 2 diabetes, managed by diet and/or metformin, were recruited via local advertisements. All participants had at least a one-year history of diabetes and a mean fasting blood glucose below 200 mg/dL, verified during a three-day prescreening period, to minimize risk during the glucose challenge.

Participants attended two laboratory sessions, separated by three days, each beginning after an overnight fast (mean fasting duration: 11.6 hours). In each session, participants consumed a beverage labeled either as "high sugar" (124 g) or "low sugar" (0g), but in reality, both beverages were identical and contained 62g of sugar. The order of beverage label presentation was counterbalanced using block randomization.

Before consuming the beverage, participants completed general baseline assessments, including measures of stress, mood, and eating behaviors. Blood glucose was measured at baseline and then at 20, 40, and 60 minutes after beverage consumption. Participants were instructed to finish the beverage within three minutes. After each session, they also rated how much sugar they believed the beverage contained and their satisfaction with the nutritional information provided on the label. This manipulation check ensured that the labels effectively shaped participants' beliefs about sugar intake.

2.3.2   Results

The label manipulation was highly effective: participants rated the "high sugar" beverage as having significantly more sugar (mean rating 4.00 out of 5) than the "low sugar" beverage (mean 1.97 out of 5), despite both drinks containing the same amount of sugar.

Crucially, blood glucose responses tracked participants' *perceptions* of sugar intake rather than the actual sugar content. Linear mixed models revealed a significant interaction between beverage label and time, indicating that blood glucose levels rose more sharply after participants *believed* they had consumed a high-sugar beverage, compared to when they believed it was low in sugar; even though both drinks were identical. This effect persisted after controlling for order of presentation and baseline glucose levels.

Further analysis showed that individual differences in eating behavior and nutritional satisfaction moderated this effect. Specifically, participants who were more influenced by external eating cues (e.g., eating in response to food-related cues in the environment, rather than internal hunger) showed greater label-driven changes in blood glucose. Mediation analysis revealed that the pathway from beverage label to blood glucose change was partially mediated by (1) perceived sugar intake and (2) satisfaction with the nutritional label. In other words, the more participants believed and felt satisfied with the label, the larger the effect on their glucose metabolism.

There were no significant effects of the manipulation on subjective stress, affect, or hunger ratings, indicating that the blood glucose changes were not simply due to psychological arousal, mood, or acute hunger.

2.3.3   Interpretation and Implications

This study provides direct evidence that the body's metabolic response (in this case, blood glucose levels in people with type 2 diabetes) can be shaped by what individuals *believe* they have consumed, rather than by the food's actual nutritional content. The findings challenge the classical biomedical assumption that glucose levels after eating a meal are determined solely by carbohydrate load and insulin action. Instead, they support a biopsychosocial model of metabolism, in which anticipation, belief, and satisfaction with nutritional information can produce real, measurable changes in core physiological processes [48].

The mechanism appears to involve "anticipatory budgeting" by the brain, in which the expectation of high sugar intake primes the body to regulate glucose accordingly, potentially via neural or hormonal pathways. The effect was strongest among individuals most attuned to external eating cues and those who were most satisfied with the label information, suggesting that psychological context and individual differences are crucial moderators.

2.3.4   Scientific Context and Critique

The Perceived Sugar Study sits within a growing literature demonstrating that mindset, expectation, and informational framing can shape physiological outcomes. Other work showed similar effects of mindset and informational context on physiological outcomes, including the Housekeepers Study (section 2.1), which showed that perceiving daily activities as exercise can improve health markers, the Perceived Time and Blood Sugar Study (section 2.2), which found that blood glucose levels follow perceived rather than actual time, and, as discussed in the next section on Mind Over Milkshakes (section 2.4), the labeling of foods can alter satiety hormone responses.

However, as with all laboratory-based studies, the generalizability of these results to broader, real-world dietary settings remains to be established. The study sample (people with relatively well-controlled type 2 diabetes) was selected for safety, which may limit applicability to those with more severe or unstable disease. Furthermore, the study did not assess the long-term durability of these effects or their impact on diabetes management outcomes over time.

Nevertheless, the implications are profound: interventions that target beliefs and expectations about food, not just its composition, may offer a new avenue for diabetes management and dietary interventions. Psychological support and nutritional education that address not only what people eat, but also what they *believe* they are eating, could play an important role in metabolic health.

> **Key Point:** Blood glucose levels after eating can be influenced as much by what you *think* you consumed (based on labels or expectations) as by what you actually consumed. This means that mindset and informational context are not just psychological, they can physically alter your body's response to food. Paying attention to how you interpret and believe in the nutritional value of your meals may be as important as the content itself.

2.4   The Mind Over Milkshakes Study

An interesting illustration of the power of mindset in shaping physiological responses to food comes from the study by Crum, Corbin, Brownell, and Salovey (2011), titled "Mind Over Milkshakes: Mindsets, Not Just Nutrients, Determine Ghrelin Response." This experiment provides direct evidence that our beliefs about what we are eating (not just the actual nutritional content) can alter the body's hormonal response to food, with important implications for appetite regulation, satiety, and potentially weight management [49].

2.4.1   Study Design and Methodology

The study recruited 46 healthy adults (aged 18–35, mean BMI 22.5, 65% women) from the Yale and New Haven community. Participants were pre-screened to exclude those with diabetes, pregnancy, chronic medical or psychiatric conditions, or allergies to milkshake ingredients. Each participant attended two laboratory sessions, scheduled one week apart, after an overnight fast.

At each session, participants were told they would be tasting a different milkshake as part of a "Shake Tasting Study" evaluating the body's response to different nutrients. In reality, both milkshakes were identical: each contained 380 calories, with the same amounts of fat, sugar, and protein. The critical manipulation was in the labeling and description of the shakes:

o **Indulgent Condition:** The shake was labeled as a "high-calorie, high-fat, indulgent" milkshake (620 calories, 30g fat, 56g sugar).

○ **Sensible Condition:** The shake was labeled as a "low-calorie, low-fat, sensible" milkshake (140 calories, 0g fat, 20g sugar).

The order of presentation was counterbalanced: about half the participants received the indulgent shake first, and half received the sensible shake first.

Upon arrival, an intravenous catheter was placed for blood draws. After a 20-minute rest, a baseline blood sample was collected. During the next 40 minutes, participants viewed and rated the shake label (anticipatory phase). At the 60-minute mark, they consumed the milkshake (within 10 minutes), after which a final blood sample was collected at 90 minutes (postconsumption phase). At each time point, participants also rated their subjective hunger and the perceived healthiness and tastiness of the shake.

The primary outcome was plasma ghrelin, a gut hormone that rises before eating (stimulating hunger) and falls after eating (signaling satiety). Ghrelin was measured at baseline, after viewing the label (anticipatory), and after consuming the shake (postconsumption). The study also assessed dietary restraint using the Dutch Eating Behavior Questionnaire.

2.4.2 Results

The label manipulation was highly effective: participants rated the "sensible" shake as significantly healthier than the "indulgent" shake, confirming that the mindset manipulation worked as intended. There were no significant differences in perceived tastiness between the two conditions.

The key finding was in the ghrelin response. When participants believed they were consuming the "indulgent" shake, their ghrelin levels rose sharply in anticipation and then dropped steeply after consumption, which is a pattern consistent with having consumed a high-calorie, satisfying meal. In contrast, when participants believed they were consuming the "sensible" shake, their ghrelin levels remained relatively flat, showing little suppression after eating, as if they had consumed a low-calorie, unsatisfying meal.

Statistical analysis revealed a significant quadratic interaction between shake label and time: the decline in ghrelin after the "indulgent" shake was much greater than after the "sensible" shake, despite both shakes being nutritionally identical. This effect was independent of the order of presentation and was not moderated by participants' level of dietary restraint.

Interestingly, subjective hunger ratings did not differ significantly between conditions, possibly due to the timing or sensitivity of the hunger measure. However, the physiological marker of satiety (ghrelin) was clearly shaped by mindset, not just by nutrients.

2.4.3   Interpretation and Implications

This study provides compelling evidence that the body's hormonal response to food is not determined solely by the actual nutrients consumed, but also by what we *believe* we are consuming. The mindset of indulgence (i.e., believing one is eating a rich, satisfying treat) produced a physiological response (a steep drop in ghrelin) consistent with greater satiety. In contrast, the mindset of restraint (i.e., believing one is eating a light, diet food) blunted the ghrelin response, potentially leaving the body in a state of continued hunger and reduced satisfaction.

These findings challenge the classical view that appetite hormones like ghrelin are regulated only by caloric intake and nutrient sensing. Instead, they support a biopsychosocial model in which beliefs, expectations, and informational context can directly influence core metabolic processes. The results suggest that approaching even healthy foods with a mindset of indulgence (allowing oneself to feel satisfied and rewarded) may enhance physiological satiety and potentially support better appetite regulation.

For dieting and weight management, this has important implications. If "diet" foods are consistently approached with a mindset of restraint or deprivation, the body may fail to register satiety, leading to increased hunger and a greater risk of overeating later. Conversely, cultivating a mindset of satisfaction and enjoyment, even with nutritious foods, may help align physiological signals with dietary goals.

2.4.4   Scientific Context and Critique

The Mind Over Milkshakes study builds on a growing literature showing that mindset and expectation can shape not only subjective experiences (such as taste and fullness) but also objective physiological markers. Other research has demonstrated similar effects for exercise (the Housekeepers Study, see section 2.1), blood glucose (Perceived Sugar Study, see sections 2.2 and 2.3), and even healing rates (Perceived Time and Healing Study, discussed in the next section 2.5). This study is unique in demonstrating that a single, subtle change in informational framing can alter the body's hormonal response to food.

As with all laboratory-based studies, questions remain about the generalizability of these effects to real-world eating patterns and long-term outcomes. The study did not assess whether the altered ghrelin response translated into differences in subsequent food intake or weight change. Future research should explore how mindset interventions can be integrated into dietary counseling and public health messaging to support sustainable eating behaviors.

Nonetheless, the implications are profound: the way we think about our food (not just what we eat) can shape our body's response at the hormonal level. Mindset matters, even for the most basic processes of hunger and satiety.

> **Key Point:** Your body's hormonal response to food is influenced not just by what you eat, but by what you *believe* you are eating. Approaching meals with a mindset of satisfaction and indulgence (even when eating healthy foods) can enhance satiety and potentially support better appetite regulation. Mindset matters, down to the level of your hunger hormones.

2.5 The Perceived Time and Healing Study

A further extension of the mind-body paradigm comes from recent research into how our subjective experience of time can directly influence physical healing. The study by Aungle and Langer (2023), titled "Physical healing as a function of perceived time," provides some evidence that the rate at which our bodies heal is not solely determined by biological processes or the objective passage of time, but can be significantly shaped by how much time we *believe* has passed [50]. This work adds a new dimension to the science of mindset, suggesting that even our most basic physiological processes are susceptible to psychological framing.

2.5.1 Study Design and Methodology

The study employed a within-subjects experimental design, recruiting 33 healthy adults (average age 28.3 years, 70% women) from a university population. The researchers used a standardized cupping procedure to create temporary, visible marks (mild wounds) on participants' forearms. Each participant completed three laboratory sessions, each on a separate day, with the cupping applied to a different location on the forearm in each session to avoid overlap.

The critical manipulation involved altering participants' *perception* of how much time had passed during the healing observation period, while keeping the actual elapsed time constant at 28 minutes in all conditions. This was achieved using a digital timer that ran at different speeds:

o **Slow Time Condition:** The timer ran at half speed, so participants believed only 14 minutes had passed.

o **Normal Time Condition:** The timer matched real time (28 minutes).

o **Fast Time Condition:** The timer ran at double speed, so participants believed 56 minutes had passed.

During each session, participants completed brief surveys every few minutes (as indicated by the timer) to rate the appearance and healing of the cupping mark, as well as their mood and stress. To ensure participants attended to the passage of time, they were instructed to monitor the timer and complete the surveys at the indicated intervals. All personal devices were collected to prevent access to real time cues.

To objectively assess healing, standardized photographs of the cupping marks were taken at the start and end of each session. These photo pairs were later rated for degree of healing by a panel of independent, blinded raters using a 10-point scale.

2.5.2   Results

The results were clear and compelling: *the more time participants believed had passed, the more their wounds appeared to heal*, even though the actual elapsed time was identical in all conditions. Specifically, mean healing ratings (on a 10-point scale) were:

o **Slow Time (perceived 14 min):** 6.17

o **Normal Time (perceived 28 min):** 6.43

o **Fast Time (perceived 56 min):** 7.30

Statistical analysis confirmed that healing was significantly greater in the Fast Time condition compared to both Normal and Slow Time, and greater in Normal compared to Slow Time. The effect was robust across participants and independent raters, and remained significant after controlling for age, stress, mood, and other psychosocial variables.

Importantly, participants' actual behavior, environment, and the physical nature of the wounds were held constant across conditions. The only difference was their *perception* of how much time had passed.

2.5.3   Interpretation and Implications

This study provides direct evidence that the body's healing processes can be modulated by psychological constructs as abstract as the perception of time. The findings support the theory of mind-body unity, which posits that mental states and expectations can exert simultaneous and bidirectional influences on physiological processes [50].

For the science of dieting and health, these results are meaningful, too. They suggest that our beliefs about time, progress, and recovery may influence not only how we feel, but how our bodies actually function. For example, if someone believes that "not enough time has passed" for a diet or exercise regimen to be effective, this mindset could potentially slow physiological changes, just as believing more time has passed can accelerate healing. Conversely, a mindset that "progress is happening quickly" may actually help the body adapt and recover more efficiently.

The study also highlights the importance of expectation and context in shaping health outcomes. Just as the perception of having exercised (Housekeepers Study) or having consumed sugar (Perceived Sugar Study) can alter physiological markers, so too can the perception of time alter the rate of physical healing. This opens new avenues for interventions that target not just behavior, but the psychological framing of time, progress, and recovery in dieting and health.

2.5.4   Scientific Context and Critique

The Perceived Time and Healing Study builds on a growing literature demonstrating that subjective experience (including beliefs about time, food, and activity) can shape objective physiological outcomes. Similar research from the same group showed that blood glucose levels in people with diabetes follow perceived time rather than actual time, and that cognitive performance after sleep deprivation is influenced by how much sleep participants *believe* they had, not how much they actually had.

As with all laboratory-based studies, questions remain about generalizability to real-world settings and long-term effects. The study's use of a within-subjects design and objective, blinded ratings of healing strengthen its conclusions, but future research is needed to explore how these effects play out in more complex, everyday health behaviors like dieting, weight loss, and exercise adherence.

Nonetheless, the implications are striking: our bodies may be more responsive to our beliefs and expectations than previously thought. For those pursuing dietary change or recovery from illness, cultivating a mindset that embraces progress and allows for the possibility of rapid, positive change may itself be a powerful tool for health.

> **Key Point:** The rate at which your body heals (and potentially adapts to diet or exercise) can be influenced by how much time you *believe* has passed, not just by the actual passage of time. This underscores the power of mindset: framing your health journey as one of steady, meaningful progress may help your body respond more effectively, while focusing on slow or insufficient change could inadvertently slow your results. Mindset matters, even for the passage of time itself.

Chapter 3

Health Claims

In today's digital age, the internet is saturated with conflicting health claims, many of which dismiss decades of careful scientific research in favor of anecdote, speculation, or sensationalism. All too often, correlation is mistaken for causation, such as when studies find that people who drink more zero-calorie sodas are also more likely to be obese. Rather than recognizing that individuals struggling with weight may choose diet sodas to reduce their sugar intake, some leap to the simplistic (and incorrect) conclusion that these sodas cause obesity.

Similar patterns appear across countless health topics: the assumption that gluten-free diets are healthier for everyone, that seed oils are inherently toxic, or that a single nutrient or food is responsible for all modern health problems. Exaggerated, absolute claims (especially those delivered without citations or grounded in oversimplified logic) can have real and sometimes dangerous consequences. Misinformation does not just confuse; it can harm or even kill. In one tragic example, an online influencer falsely claimed that type 1 diabetes could be cured by fasting, a myth that endangers lives.

As you read this chapter, remember: your health depends on conclusions drawn from facts, rigorous research, and critical thinking, not on the loudest voices or the latest viral post. Be wary of those who speak in absolutes, avoid citing credible sources, or reduce complex issues to a single nutrient or villain. This chapter is dedicated to examining real-world claims circulating online and separating fact from fiction, so you can make informed choices grounded in scientific reality.

3.1 Blending Fruit

Some popular advice suggests that blending fruit, such as making smoothies, drastically increases its effect on blood sugar; sometimes claiming it can quadruple the glycemic response (i.e., the effect it has on your blood sugar levels; specifically, how much and how quickly your blood glucose rises after eating. This is often measured using the glycemic index, or GI, which is a scale from 0 to 100 that ranks foods based on how rapidly they increase blood sugar compared to pure glucose. A higher glycemic index means a food causes a faster and larger increase in blood sugar, while a lower glycemic index means a slower and smaller rise). Hence, when people talk about a food "increasing the glycemic response," they mean it raises your blood sugar more quickly or to a higher level. Conversely, "decreasing the glycemic response" means the food leads to a slower or smaller rise in blood sugar.

For example, if blending fruit were to "quadruple the glycemic response," as some claims suggest, it would mean that your blood sugar would rise much more sharply and to a much higher level after drinking a smoothie than after eating the same fruit whole. The reasoning behind this claim is that blending supposedly disrupts the fruit's fiber structure, making sugars more rapidly available for absorption and thus causing a much higher spike in blood glucose. They further argue that this effect applies to any fruit, and that such blood sugar spikes are inherently harmful.

However, as the scientific evidence (detailed in sections below) shows, blending fruit (when all fiber is retained) does not dramatically increase the glycemic response and may even lower it for certain fruits, especially those with seeds. This is because blending can help disperse fiber, which slows down sugar absorption, resulting in a lower or more gradual increase in blood sugar compared to eating whole fruit.

> **Key Point:** To understand the following, remember that "glycemic response" is referring to how much and how quickly a food causes your blood glucose (sugar) to rise after you eat it. This is often measured using the glycemic index (GI), which is a scale from 0 to 100 that ranks foods based on how rapidly they increase blood sugar compared to pure glucose (which has a GI of 100). A higher glycemic index means the food causes a faster and larger increase in blood sugar levels. A lower glycemic index

means the food causes a slower and smaller increase in blood sugar.

3.1.1 Scientific Analysis and Rebuttal

This claim, that blending fruit is bad, does not hold up to scientific scrutiny. Controlled studies have directly compared the blood sugar response after consuming whole fruits versus their blended counterparts [51]. For example, research involving healthy young adults found that blending fruits like apples and blackberries did not increase, but actually *reduced* the postprandial (after eating) blood sugar response compared to eating the same fruits whole. The reduction was about 15 points on the blood glucose scale. The likely explanation is that blending disperses dietary fiber (especially from seeds) throughout the drink, which can slow down the absorption of sugars rather than accelerate it.

Additionally, the idea that normal, transient increases in blood sugar after eating fruit are dangerous is misleading. While chronically elevated blood sugar over many years can indeed damage organs and contribute to disease, temporary rises after meals are a normal and healthy part of metabolism. There is no evidence that blending fruit creates a harmful glycemic effect in healthy individuals.

> **Key Point:** Blending fruit does not dramatically increase its glycemic impact. In some cases, it may even lower the acute blood sugar response compared to eating whole fruit. Always check scientific evidence before accepting dietary claims.

3.1.2 Detailed Evidence: Alkutbe et al. (2020)

A particularly robust study by Alkutbe et al. (2020) directly addresses the question of whether blending fruit increases or decreases the glycemic response [52]. In this randomized crossover trial, both adults with obesity and healthy weight adults consumed two types of fruit mixtures (raspberry/mango and passionfruit/mango) in two forms: as whole fruit (chopped and chewed) and as "nutrient-extracted" (blended with water, with all fiber retained).

Each serving was standardized to contain 25g of total sugar, and blood glucose was measured at multiple intervals over two hours post-consumption. The glycemic index (GI) was calculated for each treatment.

Blending fruit *reduced* the glycemic index by 27–43% across all groups, with statistically significant reductions in obese adults. For example, in adults with obesity, the GI for raspberry/mango dropped from 44.85 (whole) to 25.43 (blended), and for passionfruit/mango from 42.56 (whole) to 26.30 (blended). Similar, though in some cases even larger, reductions were seen in healthy weight adults: for raspberry/mango, GI dropped from 68.13 (whole) to 39.89 (blended), and for passionfruit/mango from 73.49 (whole) to 30.99 (blended). The results are summarized in Table 3.1.

Table 3.1: Glycemic index (GI) values for whole versus blended fruit mixtures in adults with obesity and healthy weight adults, as reported by Alkutbe et al. (2020). Blending fruit with all fiber retained resulted in a lower GI compared to eating whole fruit.

		Glycemic Index	
Group	**Fruit Mix**	**Whole Fruit**	**Blended Fruit**
Overweight	Raspberry/Mango	44.85	25.43
Overweight	Passionfruit/Mango	42.56	26.30
Healthy Weight	Raspberry/Mango	68.13	39.89
Healthy Weight	Passionfruit/Mango	73.49	30.99

Notably, the GI values were higher in healthy weight adults than in those with obesity (see Table 3.1). The authors observed this difference but did not identify a definitive cause. However, evidence from related studies provides several plausible explanations rooted in both metabolic adaptations and methodological factors [53]. First, metabolic adaptations commonly seen in obesity (such as insulin resistance, impaired glucose tolerance, and compensatory hyperinsulinemia) can significantly alter the glycemic response to carbohydrate intake. In individuals with obesity, insulin resistance impairs the ability of tissues to take up glucose efficiently, prompting the pancreas to secrete more insulin in an attempt to maintain normal blood glucose levels [54]. Additionally, impaired glucose tolerance is a hallmark of obesity, even in young adults without overt diabetes. Obese individuals tend to have higher fasting and post-load glucose levels, but their glycemic response curves are often less pronounced due to the body's diminished ability to respond to glucose challenges [55].

The reduction in glycemic response was most pronounced for fruit blends containing seeds (raspberries, passionfruit). The authors suggest that blending disperses fiber and releases nutrients from seeds, making them more accessible and slowing sugar absorption. This is the opposite of the popular claim that blending "destroys" fiber; in fact, it may enhance the fiber's beneficial effects.

Hence, the Alkutbe et al. (2020) study provides robust, controlled evidence that blending fruit, when fiber is retained, does not increase and may even *decrease* the glycemic response compared to eating whole fruit; especially for seeded fruits. This directly contradicts the popular claim and supports the scientific rebuttal above.

> **Updated Key Point:** Blending fruit, especially when seeds and fiber are retained, does not dramatically increase its glycemic impact. In many cases, it may even lower the acute blood sugar response compared to eating whole fruit. Always check scientific evidence before accepting dietary claims.

3.1.3   Summary: Claim vs. Evidence

o **Claim:** Blending fruit (e.g., making smoothies) "quadruples" the glycemic response, causing a much sharper and higher spike in blood sugar than eating whole fruit.

o **Rationale:** Blending supposedly destroys the fruit's fiber structure, making sugars more rapidly available for absorption and thus leading to a greater and more rapid increase in blood glucose.

o **Evidence:**   In a controlled, randomized crossover trial, blending fruit **reduced** the glycemic index (GI) by 27–43% compared to eating the same fruit whole, with all fiber retained.

o **Examples:**

- **Adults with obesity:**
  Raspberry/mango GI: **44.85** (whole) → **25.43** (blended)
  Passionfruit/mango GI: **42.56** (whole) → **26.30** (blended)

- **Healthy weight adults:**
  Raspberry/mango GI: **68.13** (whole) → **39.89** (blended)
  Passionfruit/mango GI: **73.49** (whole) → **30.99** (blended)

o **Proposed Mechanism:** Blending disperses fiber (especially from seeds) throughout the drink, which can **slow down sugar absorption** rather than accelerate it.

o **Conclusion:** Blending fruit **with all fiber retained does not increase** the glycemic response and may even **lower** it, especially for seeded fruits.

– The study's results **directly contradict** the popular claim. Instead of increasing the glycemic response, blending fruit (with fiber retained) actually **decreases** it compared to eating whole fruit.

3.2 Fruit vs. Juice

While both blended fruit (such as smoothies) and fruit juice might seem similar (since both involve processing fruit into a drinkable form) their effects on blood sugar and satiety are actually quite different. This apparent contradiction arises from the crucial role of fiber: blended fruit retains all of the fruit's fiber and structure, whereas juice (especially pulp-free) removes most of it, fundamentally altering how your body processes the sugars. As a result, even though smoothies and juice may look alike, their impact on health can be dramatically different, as explained below.

There is a significant difference between eating fresh, whole fruit and drinking fruit juice, especially when the juice is pulp-free. Whole fruit contains an intact fiber matrix, which slows the absorption of sugars and provides greater satiety. In contrast, fruit juice (particularly when it is strained to remove pulp) lacks most of the fiber found in whole fruit. This difference in structure and composition leads to important distinctions in how the body processes these foods and their effects on blood sugar and appetite regulation [56,57].

3.2.1 Scientific Evidence: Satiety and Blood Sugar

Flood-Obbagy and Rolls (2009) conducted a detailed randomized crossover trial in which adults consumed apples in four forms: whole apple segments, applesauce, apple juice with added fiber, and apple juice without fiber, all matched for weight, energy, and ingestion rate [56]. Their results demonstrated that eating whole apple segments before a meal significantly reduced subsequent calorie intake and increased satiety compared to applesauce or either juice condition. Notably, even when fiber was added back to the juice, it did not replicate the satiety benefits of whole fruit, highlighting the importance of the fruit's physical structure (not just its fiber content) in promoting fullness and supporting weight management.

Chodur et al. (2020) further reinforced these conclusions by systematically reviewing clinical and epidemiological studies comparing whole fruit and 100% fruit juice [57]. Their analysis found that whole fruit, due to its intact fiber matrix, slows the absorption of sugars, resulting in a more gradual increase in blood glucose and insulin levels compared to juice. In contrast, fruit juice (especially when pulp-free) lacks most of the fiber, leading to a sharper spike in blood sugar. These metabolic differences are significant, as rapid glucose spikes are associated with increased risk for metabolic disorders and reduced satiety, which may contribute to overeating and weight gain.

3.2.2   The Role of Fiber and Pulp

The dietary fiber content of fruit juices, particularly orange juice, has been specifically investigated by Grigelmo-Miguel and colleagues [58]. Their study found that the byproducts of juice extraction, such as pulp and residues, are rich in dietary fiber; especially pectin. However, the process of juice extraction typically removes much of this fiber, especially in pulp-free juices, resulting in a beverage with significantly lower fiber content compared to whole fruit or juices with pulp. Including pulp in juice increases its fiber content and water-holding capacity, but still does not match the levels found in whole fruit.

Thus, juices that retain pulp contain more fiber than pulp-free juices, making them a somewhat better choice for those who prefer to drink juice. The presence of pulp helps slow sugar absorption and provides some of the benefits of whole fruit, though not to the same extent.

Adding nuance to this picture, Alkutbe et al. (2020) investigated the glycemic response to different forms of fruit consumption, specifically comparing whole fruit to fruit processed using a nutrient-extraction blender that retains all fiber and seeds [52]. In their randomized crossover study, they found that nutrient-extracted fruit (blended with all fiber and seeds retained) actually resulted in a significantly lower glycemic index than eating the same fruit whole, for certain fruit combinations. The authors suggest that blending may release fiber and nutrients from seeds, making them more available to slow sugar absorption. This challenges the conventional wisdom that whole fruit always produces a lower glycemic response than any processed form, at least for some fruits and preparation methods, and suggests that nutrient-extracted smoothies could have a role in diets focused on blood sugar management.

3.2.3   Summary and Practical Implications

Hence, the evidence makes it clear that not all fruit-based beverages are created equal. Whole fruit provides the best combination of fiber, nutrients, and structure to support healthy blood sugar regulation and lasting satiety. In contrast, fruit juice (especially varieties without pulp) lacks much of the fiber and structure that slow sugar absorption, leading to faster blood sugar spikes and less fullness after consumption. Interestingly, smoothies or "nutrient-extracted" blends that retain all the fruit's fiber and seeds can, in some cases, offer even better glycemic outcomes than whole fruit itself, particularly for certain seeded fruits. This highlights the importance of how fruit is processed: preserving the natural fiber matrix is key. For those who do choose juice, opting for versions with pulp is a healthier choice, but it still does not match the benefits of whole or blended fruit. Ultimately, understanding these differences can help guide healthier choices for managing blood sugar and supporting overall well-being.

> **Key Point:** Whole fruit contains more fiber and nutrients than juice, especially pulp-free juice. Eating fruit is generally better for blood sugar control and satiety than drinking juice. However, for certain fruits, nutrient-extracted smoothies that retain all fiber and seeds may produce an even lower glycemic response than whole fruit, suggesting a potential role for these preparations in diets focused on blood sugar management. If you really want juice, choose one with pulp for a healthier option.

3.3   Gluten-Free Diet

The gluten-free diet has gained substantial popularity in recent years, with widespread claims that eliminating gluten (i.e., a protein found in wheat, barley, and rye) confers broad health benefits, even for individuals without celiac disease or diagnosed gluten sensitivity. Proponents often assert that gluten is inherently inflammatory, that it impairs digestion and cognitive function, and that a gluten-free diet is generally healthier for the population at large. These claims have led to a proliferation of gluten-free products and a perception that gluten avoidance is beneficial or even necessary for optimal health.

Recent systematic reviews have specifically addressed whether gluten avoidance confers health benefits in the general population. A 2022 large-scale review found no evidence that a gluten-reduced or gluten-free diet reduces the risk of cardiovascular disease or mortality in adults without celiac disease, despite the increasing popularity of such diets in the general population [59].

Gluten is a composite of storage proteins, primarily gliadin and glutenin, found in wheat and related grains. In individuals with celiac disease, ingestion of gluten triggers an autoimmune response that damages the small intestine, leading to malabsorption and a range of clinical symptoms. Non-celiac gluten sensitivity (NCGS) is a less well-defined condition characterized by gastrointestinal and extra-intestinal symptoms related to gluten ingestion, but without the autoimmune features of celiac disease. For the vast majority of the population, however, gluten is well tolerated and does not cause adverse health effects.

> **Key Point:** A gluten-free diet is medically necessary for individuals with celiac disease and may benefit those with diagnosed non-celiac gluten sensitivity. For the general population, there is no relevant evidence that gluten is harmful or that a gluten-free diet confers health benefits.

3.3.1   Scientific Analysis and Rebuttal

The claim that a gluten-free diet is universally beneficial does not withstand scientific scrutiny. Multiple controlled studies have examined the effects of gluten consumption in individuals without celiac disease or NCGS. These studies consistently demonstrate that gluten does not cause inflammation, digestive issues, or cognitive impairment in healthy individuals. In fact, several investigations have found that gluten-free diets may be associated with nutritional deficiencies, increased intake of processed foods, and higher costs, without measurable health benefits in those without gluten-related disorders [60, 61].

For example, randomized controlled trials have shown that removing gluten from the diet of healthy adults does not improve gastrointestinal symptoms, markers of inflammation, or overall well-being. In some cases, participants on gluten-free diets reported increased fatigue and reduced intake of dietary fiber, B vitamins, and minerals, due to the exclusion of whole grains.

Furthermore, the notion that gluten is inherently inflammatory or toxic is not supported by biochemical or clinical evidence. The adverse effects of gluten are specific to individuals with celiac disease or, less commonly, NCGS. For the general population, gluten-containing foods such as whole wheat, barley, and rye are valuable sources of nutrients and dietary fiber [62].

This is supported by findings from prospective cohort studies involving hundreds of thousands of participants, which show that, over up to 26 years of follow-up, the risk of cardiovascular mortality was identical in those with low versus high gluten intake, with the risk also being essentially the same for all-cause mortality and for non-fatal myocardial infarction. Specifically, the risk of cardiovascular mortality and all-cause mortality was neither increased nor decreased with lower gluten intake, and the risk of non-fatal myocardial infarction was likewise unchanged [59]. Randomized trial evidence also indicates that a gluten-free diet does not affect blood pressure, LDL cholesterol, or BMI in healthy adults.

> **Key Point:** There is no scientific basis for recommending a gluten-free diet to the general population. Unnecessary gluten avoidance may lead to nutritional imbalances and is not associated with improved health outcomes in individuals without gluten-related disorders.

3.3.2 Detailed Evidence: Trials and Studies

A number of robust studies have directly addressed the health effects of gluten-free diets in individuals without celiac disease, focusing particularly on whether gluten itself is responsible for gastrointestinal symptoms in such populations.

One of the earliest and most influential studies was a double-blind, randomized, placebo-controlled trial by Biesiekierski et al. (2011) [63]. In this study, 34 adults with irritable bowel syndrome (IBS) who did not have celiac disease (excluded by serology and histology) and who were already symptomatically controlled on a gluten-free diet were randomized to receive either gluten (in the form of two bread slices plus one muffin per day) or placebo, while continuing their gluten-free diet for up to 6 weeks. Symptoms were assessed using a visual analog scale, and markers of intestinal inflammation, injury, and immune activation were monitored. The results showed that 68% of patients in the gluten group reported inadequate symptom control compared to 40% in the placebo group. On the visual analog scale, the gluten group experienced significantly worse overall symptoms, pain, bloating, dissatisfaction with stool consistency, and tiredness within one week. However, no mechanism for these symptoms was identified: there were no significant changes in anti-gliadin antibodies, fecal lactoferrin, celiac antibodies, C-reactive protein, or intestinal permeability. Importantly, the study also examined the role of genetic predisposition by assessing HLA-DQ2 and HLA-DQ8 status, but found no differences in response between those with or without these genetic markers.

Note on HLA-DQ2/DQ8: The human leukocyte antigen (HLA) system includes genes that play a crucial role in immune function. The HLA-DQ2 and HLA-DQ8 haplotypes are strongly associated with celiac disease, and their presence is considered necessary but not sufficient for the development of the condition. Their absence makes celiac disease highly unlikely. In studies of non-celiac gluten sensitivity (NCGS), HLA-DQ2/DQ8 status is often assessed to help exclude undiagnosed celiac disease and to explore whether genetic predisposition influences symptom response to gluten [64].

A subsequent, more rigorous double-blind, placebo-controlled, cross-over trial by the same group (Biesiekierski et al., 2013) further investigated this phenomenon [65]. In this study, 37 adults with self-reported NCGS and IBS, but not celiac disease (excluded by HLA typing or biopsy), first underwent a 2-week run-in period on a diet low in fermentable oligo-, di-, monosaccharides, and polyols (FODMAPs), which are short-chain carbohydrates known to trigger gastrointestinal symptoms. After this, participants were randomized to receive, for one week each (with washout periods in between), one of three diets: high-gluten (16,g gluten/day), low-gluten (2,g gluten/day + 14,g whey protein), or control (16,g whey protein). All meals were provided and controlled for FODMAP content.

Researchers evaluated symptoms using visual analog scales and measured both fecal markers of inflammation and indicators of immune activation. The results showed that gastrointestinal symptoms improved significantly during the low-FODMAP run-in, but worsened to a similar degree during all three dietary interventions, regardless of gluten or whey content. Only 8% of participants showed a gluten-specific effect, and no dose-response relationship was observed. There were no diet-specific changes in any biomarker, and no differences in response based on HLA-DQ2/DQ8 status. In a subsequent 3-day rechallenge phase, symptoms increased similarly across all groups (gluten, whey, or control), and gluten-specific effects could not be reproduced. An order effect was observed, with the first intervention period inducing greater symptomatic changes, regardless of content. The authors concluded that, in patients with self-reported NCGS on a low-FODMAP diet, there was no evidence of specific or dose-dependent effects of gluten.

A comprehensive review by Biesiekierski, Muir, and Gibson (2013) further contextualizes these findings, noting that while early studies suggested gluten might worsen symptoms in non-celiac individuals, more recent, rigorously controlled trials have failed to confirm that gluten is a specific trigger in patients with self-perceived NCGS. The review highlights the complexity of wheat as a food source, the potential for other components (such as FODMAPs and non-gluten proteins) to induce symptoms, and the methodological challenges in distinguishing true gluten sensitivity from other dietary intolerances. The authors emphasize that mechanisms by which gluten might trigger symptoms in non-celiac individuals have yet to be identified, and that the existence of NCGS as a discrete clinical entity remains unsubstantiated on current evidence [66].

> **Key Point:** Double-blind, placebo-controlled trials show that, when FODMAPs are controlled, gluten does not induce gastrointestinal symptoms in most individuals with self-reported non-celiac gluten sensitivity. The apparent benefit of a gluten-free diet in these individuals is likely attributable to the reduction of FODMAPs or other dietary changes, rather than gluten exclusion per se.

Large-scale population studies have also examined the long-term health outcomes associated with gluten intake. For example, a prospective cohort study involving over 100,000 participants found no association between gluten consumption and risk of coronary heart disease in individuals without celiac disease, concluding that gluten avoidance should not be recommended as a means of disease prevention in the general population [67].

Moreover, gluten-free diets are often lower in dietary fiber, iron, folate, niacin, thiamine, calcium, vitamin B_{12}, phosphorus, and zinc, due to the exclusion of fortified wheat products. This raises concerns about the potential for nutritional deficiencies, particularly when gluten-free products are highly processed and lack enrichment. These concerns are reinforced by recent evidence showing that gluten-free diets are typically lower in dietary fiber, B vitamins, and key micronutrients, and may increase the risk of nutritional deficiencies when gluten-containing whole grains are excluded without medical necessity [59].

Updated Key Point: For individuals without celiac disease or diagnosed gluten sensitivity, a gluten-free diet does not improve health outcomes and may increase the risk of nutritional deficiencies. Scientific evidence does not support the widespread adoption of gluten-free diets by the general population.

3.3.3   Summary: Claim vs. Evidence

o **Claim:** A gluten-free diet is healthier for everyone, reducing inflammation, improving digestion, and preventing chronic disease.

o **Rationale:** Gluten is purported to be inherently harmful, causing adverse effects even in those without celiac disease or gluten sensitivity.

o **Evidence:** Controlled trials and large cohort studies show **no benefit** of gluten-free diets in individuals without gluten-related disorders.

o **Examples:**

  – **Randomized controlled trials:**
    In 2011, some non-celiac IBS patients experienced more symptoms with gluten, but no mechanism was found. In 2013, when FODMAPs were controlled, no difference in symptoms or inflammation was observed between gluten and placebo diets in non-celiac adults.

  – **Population study:**
    No association between gluten intake and risk of heart disease in the general population.

o **Proposed Mechanism:** Gluten is only harmful in the context of celiac disease or NCGS; for others, it is a safe and nutritious component of the diet.

o **Conclusion:** Gluten-free diets are **not necessary** for the general population and may pose nutritional risks if not carefully managed.

  – The scientific evidence **directly contradicts** the popular claim. For most people, gluten is well tolerated, and gluten-free diets do not confer health benefits outside of specific medical indications.

3.4   Seed Oils

In recent years, seed oils (such as canola, soybean, sunflower, and safflower oils) have been recast in viral posts and podcasts as the dietary villain du jour, blamed for everything from chronic inflammation to the modern epidemic of metabolic disease. This narrative, often advanced by wellness influencers and even some credentialed professionals, claims that the widespread adoption of seed oils in the twentieth century is a root cause of rising rates of obesity, diabetes, and heart disease. The argument is seductive in its simplicity: if animal fats alone cannot explain the mid-century surge in cardiovascular mortality, then perhaps the "newcomer" seed oils are to blame.

Proponents of this view assert that seed oils are "pro-inflammatory," "toxic," or "unnatural," often invoking the language of industrial processing or chemical extraction to stoke fear. They claim that the high omega-6 fatty acid content (particularly linoleic acid) in these oils disrupts the body's omega-3 to omega-6 ratio, leading to chronic inflammation and metabolic dysfunction. Some go further, suggesting that populations with higher consumption of seed oils exhibit higher rates of heart disease and that the processing methods used to extract these oils leave harmful chemical residues.

However, as with so many dietary scapegoats, the anti–seed oil narrative is built on a foundation of selective literalism, historical misattribution, and the exploitation of epistemic voids. The scientific consensus, as articulated by major health organizations and supported by decades of peer-reviewed research, stands in stark contrast to the claims circulating online.

> **Key Point:** Seed oils are high in polyunsaturated fats, particularly linoleic acid (an omega-6 fatty acid). Understanding the scientific evidence about their health effects requires examining controlled studies and large-scale population research, rather than relying on social media claims or anecdotal reports.

3.4.1   Scientific Analysis and Rebuttal

The claims about seed oils being harmful do not hold up to scientific scrutiny. Seed oils are, in fact, high in polyunsaturated fats (particularly linoleic acid, an omega-6 fatty acid) which have been shown in large-scale cohort studies and meta-analyses to lower LDL cholesterol and reduce the risk of cardiovascular disease when they replace saturated fats in the diet.

Contrary to the assertion that seed oils are "pro-inflammatory," randomized controlled trials and systematic reviews have repeatedly found no evidence that higher intake of linoleic acid increases markers of inflammation or oxidative stress in humans. A comprehensive systematic review by Johnson and Fritsche (2012) examined 15 randomized controlled trials and found "virtually no evidence" that increasing dietary linoleic acid raises concentrations of inflammatory markers such as C-reactive protein, cytokines, or tumor necrosis factor-α in healthy adults [68]. More recent meta-analyses and umbrella reviews confirm these findings, showing that higher intake of linoleic acid or total polyunsaturated fatty acids does not increase, and may even decrease, inflammatory or oxidative stress markers in randomized controlled trials [69, 70].

On the contrary, populations with higher consumption of seed oils often exhibit lower rates of heart disease and improved metabolic profiles. The most common processing method (solvent extraction with hexane) leaves only trace residues in the final product, far below any threshold of health concern.

> **Key Point:** Controlled studies show that seed oils do not increase inflammation and may actually reduce cardiovascular disease risk when they replace saturated fats in the diet. The scientific evidence directly contradicts popular claims about their harmfulness.

3.4.2   Detailed Evidence: Summary of Studies

Multiple lines of evidence support the safety and potential benefits of seed oils when consumed as part of a balanced diet.

Meta-Analysis of Cohort Studies: A 2014 meta-analysis by Farvid et al., which pooled data from multiple prospective cohort studies, found that replacing saturated fat or carbohydrates with linoleic acid significantly reduced the risk of coronary heart disease. Randomized controlled trials included in this review demonstrated that linoleic acid intake lowers both total and LDL cholesterol levels [71].

Large-Scale Population Studies: A pooled analysis of 30 cohort studies (over 68,000 participants across 13 countries) by Marklund et al. (2019) found that higher blood levels of linoleic acid were associated with a lower risk of cardiovascular diseases, especially cardiovascular mortality and stroke [72]. Similarly, a large U.S. cohort study tracking over 200,000 people for nearly

30 years found that higher intake of plant oils (including seed oils) was linked to a lower risk of death from cardiovascular disease and cancer, while higher butter intake was linked to higher mortality [73].

Inflammation Research: The claim that seed oils are inflammatory has been extensively tested. A systematic review by Fritsche (2008) found no evidence that higher intake of linoleic acid increases markers of inflammation or oxidative stress in humans [74]. This finding has been consistently replicated in subsequent research, with studies showing that higher intake of linoleic acid or total polyunsaturated fatty acids does not increase inflammatory markers in randomized controlled trials.

International Populations: Populations that use seed oils as primary cooking fats (such as certain Mediterranean and Asian groups) consistently show lower rates of heart disease and better metabolic health compared to populations with higher saturated fat intake.

The persistence of anti–seed oil rhetoric is not the result of new scientific discoveries, but rather the product of a digital ecosystem that rewards sensationalism and oversimplification. This pattern mirrors the broader mechanisms of medical misinformation, where complex nutritional science is reduced to simple villains and heroes, stripped of scientific context.

> **Updated Key Point:** Multiple meta-analyses, large cohort studies, and randomized controlled trials consistently show that seed oils, when consumed in moderation as part of a balanced diet, are not only safe but beneficial for cardiovascular health. This conclusion is supported by major health organizations worldwide.

3.4.3 Summary: Claim vs. Evidence

○ **Claim:** Seed oils are "toxic," "pro-inflammatory," and a root cause of modern chronic diseases including obesity, diabetes, and heart disease.

○ **Rationale:** High omega-6 content supposedly disrupts the body's fatty acid balance, industrial processing leaves harmful residues, and the timing of their adoption coincides with rising disease rates.

○ **Evidence:** Large-scale studies and meta-analyses show that seed oils **reduce** cardiovascular disease risk and do **not** increase inflammation markers.

○ **Examples:**

- **Meta-analysis:** Replacing saturated fat with linoleic acid **significantly reduced** coronary heart disease risk and lowered LDL cholesterol.

- **Population studies:** Higher blood levels of linoleic acid associated with **lower** cardiovascular mortality across 68,000+ participants in 13 countries.

- **Inflammation research:** Systematic review of 15 randomized controlled trials found **"virtually no evidence"** that linoleic acid increases inflammatory markers.

- **Long-term cohort:** 30-year study of 200,000+ people found higher plant oil intake linked to **lower** mortality from cardiovascular disease and cancer.

○ **Mechanism:** Seed oils replace saturated fats, improving cholesterol profiles and reducing cardiovascular risk. Processing residues are far below safety thresholds.

○ **Conclusion:** Seed oils, when consumed in moderation as part of a balanced diet, are **safe and beneficial** for cardiovascular and metabolic health.

- The scientific evidence **directly contradicts** the popular claims. Major health organizations, including the American Heart Association, World Health Organization, and European Food Safety Authority, recommend replacing saturated fats with polyunsaturated fats from seed oils to reduce heart disease risk [75].

- The campaign against seed oils reflects the digital age's vulnerability to oversimplification and sensationalism rather than genuine medical insight.

3.5   Zero-calorie Sweeteners

In early 2023, a CNN headline made waves: "Zero-calorie sweetener linked to heart attack and stroke, study finds." The story quickly spread across social media and wellness platforms, with influencers and commentators warning their followers to avoid erythritol at all costs. The narrative was clear and alarming: erythritol, a sugar substitute found in everything from diet sodas to keto snacks, was now cast as a potential health villain responsible for cardiovascular events.

Proponents of this view assert that the study provides definitive evidence that erythritol consumption directly causes heart attacks and strokes. They claim that high levels of erythritol in the blood are a clear sign of over-consumption of artificial sweeteners, and that anyone using these products is putting themselves at serious cardiovascular risk. The argument gained traction because it seemed to offer a simple explanation for heart disease: avoid this one ingredient, and protect your cardiovascular health.

However, as with so many health scares that capture headlines and social media attention, the real story behind the erythritol study is far more complicated and nuanced than the sensationalized coverage suggests. The leap from the study's actual findings to the widespread panic about erythritol consumption represents a perfect example of how oversimplified science communication can mislead the public and create unnecessary fear.

> **Key Point:** Erythritol is a sugar alcohol used as a low-calorie sweetener, found naturally in small amounts in fruits and vegetables. Understanding the actual evidence about its safety requires examining what the studies actually measured and what they can (and cannot) tell us about causation versus correlation.

3.5.1   Scientific Analysis and Rebuttal

The claims about erythritol being a cardiovascular danger do not hold up to scientific scrutiny when the study methodology and limitations are properly understood. The study behind the headlines did not actually examine how much erythritol people were eating [76]. Instead, researchers measured the amount of erythritol circulating in people's blood and noticed that those with higher levels were more likely to have heart attacks or strokes.

This distinction is crucial because our bodies can produce erythritol naturally, especially when we're under stress or dealing with illness [77]. Erythritol is a byproduct of certain metabolic pathways that ramp up when the body is fighting inflammation or vascular damage. Therefore, high levels of erythritol in the blood might not indicate excessive sweetener consumption, but rather that the body is under metabolic strain.

Additionally, the study population was far from representative of healthy individuals. Over 70% of participants already had coronary artery disease, and the average age was 65. These were people who were already seriously ill and at high risk for heart problems regardless of their diet. When following such a high-risk population over time, cardiovascular events are almost inevitable.

The researchers also failed to account for the fact that dietary erythritol is poorly absorbed and mostly excreted in urine, so it doesn't accumulate in the body for extended periods [78]. This makes it even more likely that elevated blood erythritol levels reflect the body's own production rather than dietary intake.

> **Key Point:** The study measured erythritol levels in blood, not dietary intake, and focused on people who were already seriously ill. High blood erythritol may be a marker of existing disease rather than a cause of cardiovascular problems.

3.5.2   Detailed Evidence: Limitations and Context

The Witkowski et al. (2023) study that generated the alarming headlines has several critical limitations that were largely ignored in media coverage [76].

Study Design Flaws: The research was observational, meaning it could only show associations, not prove that erythritol causes cardiovascular events. The researchers did not assess dietary erythritol intake, instead measuring blood levels of erythritol and correlating these with cardiovascular outcomes. This is a fundamental methodological limitation because blood erythritol can come from two sources: dietary consumption or the body's own metabolic production.

Population Characteristics: The study population was heavily skewed toward individuals with existing cardiovascular disease. More than 70% of participants had established coronary artery disease, with an average age of 65 years. This represents a high-risk population where cardiovascular events would be expected to occur at elevated rates regardless of erythritol levels. The findings cannot be generalized to healthy individuals consuming erythritol-containing products.

Endogenous Production: A crucial factor overlooked in the media coverage is that erythritol is produced naturally by the human body, particularly under conditions of metabolic stress [77]. The pentose phosphate pathway, which becomes more active during oxidative stress and inflammation, can produce erythritol as a byproduct. This means that higher blood erythritol levels could be a consequence of existing cardiovascular disease rather than a cause.

Absorption and Metabolism: Dietary erythritol has unique pharmacokinetic properties that make it unlikely to accumulate in blood at the levels observed in the study [78]. Approximately 90% of consumed erythritol is absorbed in the small intestine but then rapidly excreted unchanged in urine within 24 hours. Only about 10% is metabolized by the body. This rapid clearance makes it improbable that dietary erythritol alone could account for the sustained high blood levels observed in the study participants.

Missing Confounders: The study failed to adequately control for numerous confounding factors that could explain both elevated erythritol levels and cardiovascular risk. These include diabetes status, metabolic syndrome, inflammatory conditions, and other dietary factors that might influence both erythritol production and cardiovascular outcomes.

The transformation of this complex, uncertain scientific finding into a simple "zero-calorie sweetener causes heart attacks" message represents a classic example of how nuanced research gets distorted in the translation from academic publication to public consumption.

> **Updated Key Point:** The study's design cannot establish causation, and multiple factors suggest that elevated blood erythritol is more likely a marker of existing metabolic dysfunction rather than evidence that erythritol consumption causes cardiovascular disease.

3.5.3   Summary: Claim vs. Evidence

○ **Claim:** Erythritol consumption directly causes heart attacks and strokes, and people should avoid zero-calorie sweeteners containing erythritol to protect their cardiovascular health.

○ **Rationale:** A 2023 study found that people with higher blood levels of erythritol had more cardiovascular events, supposedly proving that erythritol consumption is dangerous.

○ **Evidence:** The study **did not measure** erythritol consumption and **cannot establish** that dietary erythritol causes cardiovascular disease.

○ **Examples:**

- **Study population:** Over **70% already had coronary artery disease** with average age of 65 years, i.e., a high-risk group where cardiovascular events are expected regardless of diet.

- **Methodology:** Researchers measured **blood erythritol levels, not dietary intake**, ignoring that the body produces erythritol naturally during metabolic stress.

- **Metabolism:** Dietary erythritol is **90% excreted in urine within 24 hours**, making it unlikely to accumulate at the levels observed in sick patients.

- **Causation:** High blood erythritol may be a **marker of existing disease**, not a cause of cardiovascular problems.

o **Proposed Mechanism:** Elevated blood erythritol likely reflects the body's own production during metabolic stress and inflammation, rather than dietary consumption.

o **Conclusion:** The study **does not support** the claim that erythritol consumption causes cardiovascular disease.

- The media coverage **misrepresented** the study's findings by transforming "erythritol in blood is associated with heart problems in sick people" into "zero-calorie sweetener causes heart attacks."

- This represents a **dangerous leap** from correlation to causation that exploits public trust and spreads unnecessary fear about a ingredient that may be completely unrelated to the observed health outcomes.

3.6   The "Animal Fat Paradox"

A recurrent viral claim, sometimes echoed by credentialed physicians, asserts that early 20th-century Americans consumed large quantities of animal fats while experiencing "very low" rates of cardiovascular disease. This argument is often invoked to question the scientific consensus on the risks of saturated fats and to instead blame seed oils or other modern dietary trends for the rise in heart disease. The narrative suggests: if animal fats were truly to blame, why were rates of cardiovascular disease so much lower in 1900, when such fats predominated in the American diet?

3.6.1   Scientific Analysis and Rebuttal

This line of reasoning exemplifies a classic case of historical misattribution and the dangers of ignoring confounding variables. The reality, as illuminated by epidemiological and demographic research, is that the United States in 1900 was demographically and medically unrecognizable compared to today. Life expectancy hovered around 48 years, i.e., well below the age range (typically 65 and older) in which most cardiovascular events occur. As a result, the majority of Americans simply did not live long enough to develop or die from ischemic heart disease. Using these early 20th-century numbers as evidence of dietary safety disregards the primary risk factor for almost all chronic diseases: age.

Furthermore, diagnostic technology at the time was profoundly limited. The tools now considered standard (electrocardiograms, angiography, blood biomarkers for cardiac events) were either unavailable or in their infancy. The development and widespread adoption of these technologies in the mid-twentieth century led to increased awareness, detection, and reporting of cardiovascular disease, further distorting any direct historical comparison. Before that, many deaths from heart attacks would have been unrecognized, misclassified, or simply accepted without further investigation. Death certificates rarely listed myocardial infarction, especially since autopsies were rare outside major city hospitals. As in other domains of medicine, the absence of a diagnosis in historical records is more often a reflection of limited means for detection than of genuine rarity. Thus, the seemingly low incidence of heart disease was, in large part, a failure of detection rather than a triumph of dietary virtue.

> **Key Point:** Apparent low rates of cardiovascular disease in early 20th-century America reflect short life expectancy and limited diagnostic capabilities, not evidence that high animal fat intake was harmless.

3.6.2   Detailed Evidence: Historical and Epidemiological Data

The trajectory of cardiovascular disease mortality in the postwar era is equally significant. Contrary to the prevailing viral narrative, the United States saw a marked increase in age-adjusted cardiovascular mortality rates through the mid-20th century, peaking around 1968, followed by a remarkable decline of nearly 70% since the 1950s [79–82]. This substantial decrease corresponded with the advent of public health initiatives, advancements in acute medical care, the widespread use of statin therapy (i.e., cholesterol-lowering drugs), and, notably, changes in dietary fat consumption. However, the connection between specific types of dietary fats and disease outcomes remains intricate and influenced by multiple factors.

The popularity of the "animal fat paradox" in viral posts and podcasts, and its endorsement by some medical professionals, highlights a persistent pattern: even experts can be drawn to tidy historical narratives that align with pre-existing beliefs or modern dietary trends. The willingness to overlook demographic, medical, and technological context in favor of a seductive story is not unique to the lay public; it is a common human vulnerability. This case demonstrates not only the dangers of oversimplifying nutritional science, but also the broader risk of cherry-picking historical data to support modern agendas. It serves as a reminder that the authority of the speaker does not guarantee the validity of the claim, and that history, when stripped of context, becomes a weapon for misinformation rather than a guide to truth.

3.6.3   Summary: Claim vs. Evidence

o **Claim:** Early 20th-century Americans ate large amounts of animal fat but had very low rates of cardiovascular disease, suggesting animal fats are not to blame for heart disease.

o **Rationale:** If animal fats were truly harmful, heart disease rates should have been high when animal fat intake was high.

o **Evidence:**

  – Life expectancy in 1900 was only about 48 years, so most people did not live long enough to develop cardiovascular disease.

  – Diagnostic tools for heart disease were unavailable or primitive, leading to underdiagnosis and misclassification of deaths.

  – Age-adjusted cardiovascular mortality actually **rose** through the mid-20th century, peaking in the late 1960s, then declined by 70% with public health interventions, improved care, and dietary changes [79–82].

○ **Mechanism:** The apparent "paradox" is explained by demographic and technological factors, not by the harmlessness of animal fats.

○ **Conclusion:** The "animal fat paradox" is a misleading narrative that ignores critical context. Historical data do not support the claim that high animal fat intake is safe; rather, they reflect differences in life expectancy, diagnostic capability, and public health progress.

　　– The scientific and historical evidence directly contradicts the popular claim. Apparent low rates of heart disease in the early 1900s were due to short lifespans and poor detection, not dietary virtue.

3.7 Genetically Modified Organisms

A common claim is that genetically modified organisms (GMOs) are "unnatural" and inherently unsafe to eat or harmful to the environment. This narrative is fueled by images of scientists in labs creating "Frankenfoods," and is often reinforced by conspiracy theories and sensational media coverage. Proponents argue that GMOs introduce foreign genes into crops in ways that could not occur in nature, and that these modifications may pose unknown risks to human health and the environment. As a result, some advocate for strict regulation or outright bans on GMOs, asserting that they are fundamentally different from conventionally bred crops.

3.7.1 Scientific Analysis and Rebuttal

This claim does not withstand scientific scrutiny. The process of modifying the genetic makeup of crops is not new; humans have been altering plant genomes for thousands of years through selective breeding. Modern genetic engineering is simply a more precise and efficient extension of these traditional methods. While selective breeding relies on naturally occurring genetic variation and takes many generations, genetic engineering allows scientists to introduce specific genes responsible for desirable traits directly into a plant's genome. This enables the development of crops with improved yield, pest resistance, or nutritional content in a fraction of the time.

Extensive research and regulatory oversight have consistently shown that GMOs are as safe as their non-GMO counterparts. Major scientific organizations (including the World Health Organization, American Medical Association, and European Food Safety Authority) have all concluded that

GMOs pose no greater risk to human health than conventionally bred crops [83]. Furthermore, many GMO crops are designed to reduce the environmental impact of agriculture, such as by decreasing the need for chemical pesticides or improving drought resistance [84, 85].

> **Key Point:** Modern GMOs use targeted genetic engineering to achieve goals similar to those of traditional breeding, but with greater precision. Decades of research and regulatory review show that GMOs are as safe as conventional crops for both human health and the environment.

3.7.2   Detailed Evidence: Safety and Impact

Human Health: A comprehensive review by Goodman et al. (2024) summarizes decades of research and regulatory assessments, concluding that GMOs currently on the market are as safe to eat as their non-GMO counterparts [83]. Multiple meta-analyses and systematic reviews have found no credible evidence that GMOs cause allergies, toxicity, or other health problems in humans. Regulatory agencies worldwide require rigorous safety testing before approving GMO crops for commercial use, including assessments of potential allergenicity, toxicity, and nutritional equivalence.

Environmental Impact: A systematic review by Nicolia et al. (2013) analyzed over 1,700 studies on the environmental and health impacts of GMOs, finding no substantiated evidence of harm [84]. Many GMO crops, such as Bt corn and Bt cotton, contain genes from the bacterium *Bacillus thuringiensis* (Bt) that produce proteins toxic to specific pests. This reduces the need for chemical pesticides, which can harm beneficial insects and pollute waterways. Other GMOs, like Golden Rice, are engineered to address micronutrient deficiencies (e.g., vitamin A) in populations that rely on staple crops. Recent research also highlights the potential of GMOs to improve drought resistance and reduce greenhouse gas emissions by enabling more efficient agricultural practices [85].

Regulation and Public Perception: Despite the scientific consensus, public skepticism remains high, particularly in regions like Europe where GMOs face strict regulatory hurdles. These regulations are often implemented to address public fears, but evidence suggests that regulatory measures alone are insufficient to change deeply held beliefs. Transparent communication and education are essential to foster a more informed public discourse.

Updated Key Point: Decades of research and regulatory review show that GMOs are as safe as conventional crops for human health and the environment. Many GMOs offer environmental benefits, such as reduced pesticide use and improved resilience to climate stress.

3.7.3 Summary: Claim vs. Evidence

○ **Claim:** GMOs are unnatural, unsafe to eat, and harmful to the environment.

○ **Rationale:** Genetic engineering introduces foreign genes in ways that do not occur in nature, potentially creating unknown risks.

○ **Evidence:** Decades of research and regulatory review show that GMOs are as safe as conventional crops for both human health and the environment.

○ **Examples:**

    – **Human health:** No credible evidence of increased allergenicity, toxicity, or other health risks from approved GMOs [83].

    – **Environmental impact:** GMO crops like Bt corn and cotton reduce pesticide use and environmental pollution [84].

    – **Nutritional benefits:** Golden Rice is engineered to address vitamin A deficiency in populations relying on rice as a staple food.

    – **Climate resilience:** Drought-resistant GMOs help conserve water and reduce greenhouse gas emissions [85].

○ **Proposed Mechanism:** Genetic engineering is a precise extension of traditional breeding, enabling targeted improvements without introducing new categories of risk.

○ **Conclusion:** The scientific consensus is clear: GMOs are as safe as conventional crops and can provide significant environmental and nutritional benefits. Public skepticism is driven more by misinformation and fear than by scientific evidence.

    – Regulatory measures and transparent communication are important, but addressing the root causes of public skepticism requires ongoing education and engagement.

3.8   The Iodized Salt Controversy

A growing claim promoted by modern wellness influencers asserts that iodized salt is "unnatural," "chemically contaminated," and harmful to health. These proponents argue that iodine fortification, implemented in the 1920s, represents unnecessary "chemical processing" and advocate for expensive alternatives like Himalayan pink salt, Celtic salt, or sea salt as "pure," "natural" options. They claim these alternatives are inherently cleaner, safer, and provide superior health benefits compared to conventional iodized table salt. Some extreme voices, like self-styled naturopaths, even claim that table salt is "dangerous" because it contains sodium and chloride, asserting that these components would be lethal if injected into the blood, while promoting specialty salts with "82 minerals" as safer alternatives for conditions like high blood pressure.

The reasoning behind these claims centers on the belief that any form of food processing or fortification is inherently harmful, that "natural" automatically means better, and that the trace minerals found in alternative salts provide significant health advantages. This narrative often frames iodine fortification as a form of "corporate contamination" imposed by "Big Food" rather than recognizing it as a vital public health intervention.

> **Key Point:** Iodine fortification of salt began in the 1920s as a public health measure to prevent goiter and other iodine deficiency disorders. Understanding the actual evidence requires examining both the benefits of iodization and the real composition and safety profiles of alternative salts.

3.8.1   Scientific Analysis and Rebuttal

These claims about iodized salt being harmful do not hold up to scientific scrutiny. Iodine fortification represents one of the most successful public health interventions in history, virtually eliminating goiter (thyroid gland enlargement caused by iodine deficiency) and other iodine deficiency disorders in populations where it has been implemented. The addition of iodine to salt is a precisely controlled process that provides an essential nutrient many people would otherwise lack in their diets.

Contrary to claims that "natural" salts are inherently superior, laboratory analyses reveal that minimally processed alternative salts frequently contain detectable levels of heavy metals, including lead, arsenic, cadmium, and aluminum. In contrast, highly processed iodized salts generally contain the lowest concentrations of these contaminants, often below detectable thresholds. This demonstrates that industrial processing and purification steps actually remove harmful impurities rather than introduce them.

The assertion that table salt is dangerous because it contains sodium and chloride reveals a fundamental misunderstanding of basic physiology. Sodium chloride (identical to table salt) is the main ingredient in intravenous saline solutions administered safely to millions of patients daily in hospitals worldwide. These "dangerous" ions are actually essential for maintaining proper fluid balance, nerve function, and cellular processes.

Clinical evidence directly contradicts claims of superior health benefits from alternative salts. Studies comparing Himalayan salt to conventional table salt find no significant differences in their physiological effects, while consumers who avoid iodized salt face real risks of developing preventable thyroid conditions.

> **Key Point:** Scientific evidence shows that iodized salt is safe and provides essential nutrition, while "natural" alternatives often contain more contaminants and lack crucial iodine fortification. The processing of iodized salt removes harmful impurities rather than introducing them.

3.8.2   Detailed Evidence: Studies and Findings

Multiple lines of evidence demonstrate both the safety and necessity of iodized salt, as well as the potential risks of alternative salts.

Thyroid Health Outcomes: A large study of almost 10,000 adults found that those who avoided iodized salt were approximately 36% more likely to develop thyroid nodules compared to those who used iodized salt [86]. This demonstrates the real health consequences of avoiding iodine fortification in favor of unfortified alternatives.

Heavy Metal Contamination: Laboratory analyses by Mamavation, an independent consumer advocacy group, tested 23 popular salt products at an EPA-certified facility and found that 100% of alternative salts contained measurable amounts of heavy metals. Certain brands of Himalayan salt showed particularly high levels, with over 250 parts per million (ppm) of aluminum and up to 553 parts per billion (ppb) of lead. In contrast, highly processed iodized salts generally contained the lowest concentrations of heavy metals, often below detectable thresholds.

Clinical Comparison Studies: A randomized controlled trial published in 2022 directly compared Himalayan salt and common table salt in hypertensive individuals, finding no significant differences between the two salts in terms of their effects on blood pressure or urinary sodium concentration [87]. This directly undermines marketing claims that Himalayan salt provides superior health benefits.

Iodine Deficiency Research: Studies in regions where consumers have switched to trendy alternatives like Himalayan pink salt (often marketed at premium prices up to 20 times higher than iodized table salt) show increased risk of developing preventable thyroid conditions [88]. The trace minerals in these alternative salts provide negligible health benefits, as achieving any meaningful physiological effect would require consuming dangerously high quantities that could lead to sodium toxicity.

The persistence of anti-iodization narratives demonstrates how profit-driven actors can exploit public health initiatives by weaponizing distrust and marketing unnecessary alternatives. Social media has accelerated this dynamic, with wellness influencers generating significant revenue through affiliate links to specialty salt products while amplifying unfounded fears about iodization.

> **Updated Key Point:** Laboratory testing shows that "natural" salts often contain more contaminants than processed iodized salt, while providing no health advantages and depriving consumers of essential iodine. To obtain any meaningful health benefits from the trace minerals found in alternative salts, one would need to consume over 3 pounds of salt, which would cause dangerous sodium toxicity long before achieving any nutritional benefit. The scientific evidence strongly supports the safety and necessity of iodine fortification.

3.8.3   Summary: Claim vs. Evidence

○ **Claim:** Iodized salt is "unnatural," "chemically contaminated," and harmful, while alternative salts like Himalayan pink salt are "pure," "natural," and provide superior health benefits.

○ **Rationale:** Any food processing or fortification is inherently harmful; "natural" products are automatically better; trace minerals in specialty salts provide significant health advantages.

○ **Evidence:** Laboratory testing and clinical studies show that iodized salt is **safer and more beneficial** than alternative salts.

○ **Examples:**

– **Thyroid health:** People avoiding iodized salt were **36% more likely** to develop thyroid nodules [86].

– **Contamination: 100% of alternative salts** contained heavy metals; Himalayan salt showed over 250 ppm aluminum and up to 553 ppb lead, while processed iodized salts had levels **below detectable thresholds.**

– **Clinical outcomes:** Randomized trial found **no significant differences** in blood pressure effects between Himalayan salt and table salt [87].

– **Public health impact:** Iodine fortification virtually **eliminated goiter** and other iodine deficiency disorders where implemented.

○ **Proposed Mechanism:** Industrial processing removes harmful contaminants and adds essential iodine; "natural" salts retain environmental pollutants while lacking crucial nutrients.

○ **Conclusion:** Iodized salt is **safer, cleaner, and more nutritious** than expensive alternatives marketed as "natural."

– The scientific evidence **directly contradicts** the wellness industry claims. Consumers switching to alternative salts pay premium prices (up to 20 times higher) for products that increase contaminant exposure while eliminating essential nutrition.

– This case exemplifies how **profit-driven misinformation** can transform successful public health interventions into opportunities for predatory marketing, demonstrating the dangerous consequences when wellness narratives override scientific evidence.

3.9   HDL/LDL Cholesterol

Before getting into the details of cholesterol and its impact on heart health, it is useful to understand the basics of lipid biology. Lipids are a broad group of naturally occurring molecules that include fats, oils, and cholesterol. They are vital for life, serving as building blocks for cell membranes, storing energy, and helping the body produce important hormones. However, lipids do not dissolve in water; they are water-insoluble. Since our blood is mostly water, this presents a challenge: if lipids were simply released into the bloodstream, they would clump together, much like oil does when mixed with water, and would not be able to circulate efficiently or safely throughout the body. To overcome this, the body packages lipids into tiny particles called lipoproteins. These particles have an outer surface that interacts well with water, allowing them to move smoothly through the bloodstream. In this way, lipoproteins act like delivery trucks, carrying cholesterol and other fats to where they are needed; or, in some cases, to places where they can cause harm.

In the realm of cardiovascular health, few topics have generated as much confusion and misinformation as cholesterol; specifically, the roles of high-density lipoprotein (HDL, often called "good" cholesterol) and low-density lipoprotein (LDL, or "bad" cholesterol). Online, a variety of sensational claims circulate, ranging from the assertion that "high cholesterol increases longevity" to the idea that "all cholesterol is good for you" and that "statins are useless." These narratives, often amplified by social media and alternative health influencers, directly contradict decades of rigorous scientific research and can have dangerous consequences for public health.

3.9.1   Popular Claims and Their Rationale

o **Claim 1:** "High cholesterol increases longevity." Viral posts and articles have cited a flawed study to argue that people with high cholesterol live longer, suggesting that efforts to lower cholesterol are misguided or even harmful [89].

o **Claim 2:** "Statins are useless and the benefits of lowering cholesterol are exaggerated." Some sources claim that statins, the main class of cholesterol-lowering drugs, are ineffective and that the risks of high LDL cholesterol are overstated.

o **Claim 3:** "Cholesterol testing is a scam." Certain online forums and influencers suggest that cholesterol screening is unnecessary, designed only to sell medications or tests.

o **Claim 4:** "All cholesterol is good for you." Alternative health websites sometimes claim that the distinction between "good" (HDL) and "bad" (LDL) cholesterol is a myth, and that all cholesterol is beneficial.

○ **Claim 5:** "You don't need to worry about cholesterol if you're young." Some argue that cholesterol is only a concern for older adults, ignoring evidence that risk accumulates over time.

These claims are often rooted in misinterpretations of epidemiological data, selective citation of outlier studies, or a misunderstanding of basic lipid biology. They are further fueled by distrust of pharmaceutical interventions and a preference for "natural" approaches, regardless of scientific validity.

> **Key Point:** Outrageous claims about cholesterol (such as "high cholesterol increases longevity" or "statins are useless") are not supported by the scientific consensus. These narratives misrepresent decades of research and can lead to harmful health decisions.

3.9.2   Scientific Analysis and Rebuttal

Claim 1: "High cholesterol increases longevity." This assertion is directly contradicted by a vast body of evidence demonstrating that high LDL cholesterol is a major risk factor for atherosclerotic cardiovascular disease, the leading cause of death worldwide [90]. The Framingham Heart Study and numerous subsequent cohort studies have established a clear, dose-dependent relationship between elevated LDL and increased risk of coronary artery disease, stroke, and other vascular events [90–94]. While some flawed or methodologically limited studies have suggested paradoxical associations in select elderly populations [89], these findings are not generalizable and are often explained by confounding factors such as reverse causation, survivor bias, or failure to account for statin use.

Claim 2: "Statins are useless and the benefits of lowering cholesterol are exaggerated."

This claim ignores the extensive evidence from large, well-designed studies showing that statins lower the risk of heart attacks, strokes, and death from heart disease in both people who have already had heart problems and those at risk of developing them [90, 95–97].

Statins lower "bad" cholesterol (LDL) in the blood by blocking a key step that the liver uses to make cholesterol. When this happens, the liver pulls more cholesterol out of the blood, which lowers overall cholesterol levels. More specifically, statins mainly work by blocking a natural chemical (an enzyme) that the liver needs to make cholesterol. When the liver can't make as much cholesterol, it removes more cholesterol from the bloodstream, so blood cholesterol goes down [90].

The greater the reduction in LDL cholesterol, the greater the benefit; meaning that more intensive statin or cholesterol-lowering therapy leads to a bigger drop in the risk of heart attacks and strokes [90,96,98]. Statins are also very safe, and while they may cause a small increase in "good" cholesterol (HDL), their main and most important effect is lowering LDL [90].

Claim 3: "Cholesterol testing is a scam." Cholesterol testing is a validated, essential tool for identifying individuals at risk for atherosclerotic cardiovascular disease and guiding evidence-based treatment decisions [90]. Guidelines from major health organizations recommend routine lipid screening for adults, especially those with risk factors such as hypertension, diabetes, or family history of early heart disease [92].

Claim 4: "All cholesterol is good for you." This misrepresents basic lipid biology. LDL cholesterol is the primary driver of atherosclerotic plaque formation, while HDL cholesterol facilitates reverse cholesterol transport and has anti-inflammatory and antioxidant properties [90,99,100]. The distinction between "good" and "bad" cholesterol is not arbitrary; it is grounded in decades of mechanistic, epidemiological, and clinical trial evidence.

Claim 5: "You don't need to worry about cholesterol if you're young." Longitudinal studies show that cholesterol levels in youth and young adulthood predict cardiovascular risk later in life [92]. Early intervention can prevent the cumulative vascular damage that leads to clinical events in middle and older age.

> **Key Point:** The scientific consensus is unequivocal: high LDL cholesterol increases cardiovascular risk, statins are effective and safe for appropriate patients, and cholesterol testing is a cornerstone of preventive medicine. The distinction between HDL and LDL is biologically meaningful and clinically relevant.

3.9.3   Detailed Evidence: Summary of Studies and Data

Epidemiological Foundations: The Framingham Heart Study, launched in 1948, was pivotal in establishing cholesterol as a major risk factor for coronary artery disease [91, 92]. Subsequent studies confirmed that both dietary and cholesterol levels are associated with increased incidence of atherosclerotic heart disease [93, 94].

Randomized Controlled Trials: The CARE trial (Sacks et al., 1996) randomized post-MI patients to pravastatin or placebo, showing significant reductions in coronary events and stroke with statin therapy [95]. The LIPID trial (1998) demonstrated reduced mortality with pravastatin in a broad range of cholesterol levels. The PROVE-IT and IMPROVE-IT studies showed that more intensive LDL lowering (with higher-dose statins or statin plus ezetimibe) led to further reductions in cardiovascular events [96, 98]. Novel agents have enabled even lower LDL targets, with additional benefit and no significant increase in adverse effects [97, 101].

HDL and Cardiovascular Risk: Low HDL is an independent risk factor for coronary artery disease, but pharmacologic attempts to raise HDL have not translated into clinical benefit when LDL is well controlled [90, 102–104]. The main value of HDL appears to be in its role in reverse cholesterol transport and its protective effects on the endothelium, especially in the context of elevated LDL [99, 100].

Lifestyle and Non-Pharmacologic Interventions: Lifestyle modification (including diet, exercise, and smoking cessation) remains foundational for managing cholesterol problems. Regular endurance exercise increases HDL and improves its functionality, while smoking cessation can raise HDL by 5–10% [90, 105–107].

Genetic and Racial Differences: Rare genetic disorders such as Tangier disease (very low HDL and LDL) are not associated with premature atherosclerosis, highlighting the importance of LDL as the primary driver of risk [90, 108]. Racial differences in the impact of HDL on risk have also been observed [109].

> **Key Point:** Large-scale trials and decades of research confirm that lowering LDL reduces cardiovascular events and mortality. Raising HDL pharmacologically has not shown benefit when LDL is controlled, but lifestyle measures remain important.

3.9.4 Summary: Claim vs. Evidence

o **Claim:**

– High cholesterol (including high LDL) increases longevity and is not a health concern.

– Statins are ineffective or unnecessary for cardiovascular risk reduction.

– Cholesterol testing is a scam or unnecessary.

- All cholesterol is beneficial, and the distinction between "good" (HDL) and "bad" (LDL) cholesterol is a myth.

- Cholesterol is only a concern for older adults; young people need not worry.

○ **Rationale:**

- These claims are often based on misinterpretation of flawed or selective studies, distrust of mainstream medicine, and a preference for "natural" approaches over evidence-based interventions.

○ **Evidence:**

- Decades of rigorous epidemiological and clinical research have established that high LDL cholesterol is a major, modifiable risk factor for atherosclerotic cardiovascular disease, the leading cause of death worldwide [90–92, 94].

- Large randomized controlled trials demonstrate that statins and other LDL-lowering therapies significantly reduce the risk of heart attacks, strokes, and cardiovascular mortality in both primary and secondary prevention [90, 95–98].

- Cholesterol testing is a validated tool for identifying individuals at risk and guiding treatment; it is recommended by all major health organizations [90, 92].

- The distinction between LDL ("bad") and HDL ("good") cholesterol is biologically and clinically meaningful: LDL promotes atherosclerotic plaque formation, while HDL facilitates reverse cholesterol transport and has anti-inflammatory and antioxidant properties [90, 99, 100].

- Longitudinal studies show that cholesterol levels in youth and young adulthood predict cardiovascular risk later in life, and early intervention can prevent cumulative vascular damage [92].

- Rare genetic disorders with low LDL and HDL (e.g., Tangier disease) are not associated with premature atherosclerosis, underscoring the central role of LDL in risk [90, 108].

○ **Proposed Mechanism:**

- LDL cholesterol is the primary driver of atherosclerotic plaque formation and vascular inflammation.

- HDL cholesterol promotes removal of LDL from vessel walls and provides additional vascular protection.

○ **Conclusion:**

- The scientific consensus is unequivocal: high LDL cholesterol increases cardiovascular risk, statins are effective and safe for appropriate patients, and cholesterol testing is essential for prevention.

- The distinction between HDL and LDL is real and clinically relevant; ignoring LDL levels or dismissing statin therapy increases the risk of heart disease and stroke.

- Outrageous online claims about cholesterol are not just misleading—they are potentially harmful. Evidence-based management of cholesterol saves lives.

Key Point: High LDL cholesterol is a major risk factor for cardiovascular disease; statins and other therapies reduce this risk. Claims that "high cholesterol increases longevity" or that "statins are useless" are thoroughly debunked by decades of research and large clinical trials. Always rely on scientific evidence, not viral posts, when making decisions about cholesterol and heart health.

3.10   Egg Consumption

A widely circulated claim, originating from NutritionFacts.org (one of the largest vegan health platforms on the internet) asserts that consuming even one egg per week may increase the risk of type 2 diabetes by as much as 76%. This assertion is often presented as a universal health warning, implying that even minimal egg intake poses significant metabolic risks for all individuals, regardless of context or population.

3.10.1   Scientific Analysis and Rebuttal

The claim that one egg per week substantially increases diabetes risk is not supported by the scientific literature. The figure of a 76–77% increased risk is not derived from studies comparing minimal egg consumption (such as one egg per week) to abstention, but rather from analyses contrasting the highest levels of intake (typically one or more eggs per day) with the lowest (less than one egg per week) [110]. This distinction is critical, as it demonstrates that the claim exaggerates the risk associated with low or moderate egg consumption.

Moreover, the increased risk of diabetes associated with egg intake has been observed exclusively in U.S. populations, with no significant associations found in European or Asian cohorts [111]. This geographic specificity is often overlooked in generalized warnings, which fail to account for substantial differences in dietary patterns and lifestyle factors across regions. The context in which eggs are consumed (often alongside processed meats and refined carbohydrates in Western diets) may confound the observed associations [112]. When studies adjust for these confounding variables, the association between egg intake and diabetes risk is substantially weakened or eliminated.

> **Key Point:** The claim misrepresents the dose-response relationship and ignores geographic and dietary context. The increased risk is seen only at high levels of egg consumption in U.S. populations, and is likely confounded by broader dietary patterns.

3.10.2   Detailed Evidence: Studies and Findings

Meta-Analytical Findings: Meta-analyses indicate that in U.S. populations, each additional egg consumed per day is associated with a 14–18% higher risk of type 2 diabetes, but this association does not extend to other global populations [110, 111]. The relationship between egg consumption and diabetes risk is non-linear and appears to be highly dependent on the broader dietary context, suggesting that eggs alone are unlikely to be the primary driver of increased risk. In Western dietary patterns, eggs are frequently consumed alongside foods such as processed meats and refined carbohydrates, which themselves are linked to adverse metabolic outcomes [110]. The modest caloric contribution of a single egg (approximately 66–78 kcal) further challenges the plausibility of significant metabolic harm from moderate consumption.

Adjustment for Confounders: When studies control for dietary quality, physical activity, and body mass index, the association between egg intake and diabetes risk is substantially weakened or eliminated [112].

Randomized Controlled Trials: Evidence from randomized controlled trials supports this conclusion: interventions allowing up to two eggs per day in individuals with pre-diabetes or established type 2 diabetes show no adverse effects on glycemic control, insulin sensitivity, or other metabolic parameters [113]. In some cases, moderate egg consumption is associated with improvements in HDL and other markers of metabolic health.

In healthy individuals, a large meta-analysis of randomized controlled trials found that higher egg consumption (typically 1–3 eggs per day) leads to a modest increase in LDL (mean difference ∼8 mg/dL) and the LDL/HDL ratio (mean difference 0.14), but does not significantly affect HDL at lower intakes (e.g., one egg per day) [114]. Notably, the increase in LDL/HDL ratio is small relative to clinical thresholds for cardiovascular risk, and the effect on HDL is minimal or even positive at higher intakes. These findings suggest that, in healthy populations, moderate egg consumption does not produce adverse changes in lipid profiles likely to drive diabetes or cardiovascular risk. The authors conclude that individuals with borderline high LDL or LDL/HDL ratios may wish to limit high egg intakes, but for most healthy people, moderate consumption is unlikely to be harmful [114].

Nutritional Composition: Eggs are notable for their high-quality protein, essential vitamins (B12, A, D), minerals (selenium, iron), choline, and antioxidants (lutein, zeaxanthin), while providing relatively little saturated fat compared to other animal protein sources.

Protein Substitution Studies: Replacing animal proteins, including eggs, with plant proteins is associated with reductions in all-cause and cardiovascular mortality, with the most pronounced effects observed when plant proteins replace red and processed meats [115]. However, the magnitude of benefit is less striking when eggs are specifically replaced, underscoring the importance of considering the overall dietary pattern.

> **Updated Key Point:** Randomized trials and meta-analyses do not support a causal relationship between moderate egg consumption and diabetes risk. Observed associations in U.S. populations are likely due to confounding by dietary patterns. Randomized controlled trials show only modest, clinically insignificant changes in lipid profiles with moderate egg intake in healthy individuals [114].

3.10.3   Summary: Claim vs. Evidence

o **Claim:** Consuming even one egg per week increases the risk of type 2 diabetes by 76%, implying that even minimal egg intake poses significant metabolic risk for all individuals.

o **Rationale:** The claim, popularized by NutritionFacts.org, is based on the belief that eggs are inherently harmful and that any amount, even as little as one per week, can substantially increase diabetes risk.

o **Evidence:** Scientific analysis shows that the 76% risk figure is derived from comparisons of high (one or more eggs per day) versus very low (less than one egg per week) consumption, not from minimal weekly intake. Increased diabetes risk is observed only in U.S. populations and is not seen in European or Asian cohorts [110, 111].

o **Examples:**

  – **Meta-analyses:** Each additional egg per day is associated with a 14–18% increased diabetes risk in U.S. populations, but no significant association is found in European or Asian populations [110, 111].

  – **Confounding:** In the U.S., eggs are often consumed with processed meats and refined carbohydrates, confounding the association with diabetes risk [112].

  – **Intervention studies:** Randomized controlled trials allowing up to two eggs daily in people with pre-diabetes or diabetes show no adverse effects on glycemic control or metabolic health; some report improvements in HDL cholesterol [113]. In healthy individuals, RCTs show that moderate egg consumption (1–2 eggs/day) produces only a small increase in LDL/HDL ratio (mean difference 0.14) and LDL (mean difference ~ 8 mg/dL), with no significant effect on HDL [114].

  – **Protein substitution:** Replacing animal proteins (including eggs) with plant proteins reduces mortality risk, but the greatest benefit is seen when replacing red and processed meats [115].

o **Proposed Mechanism:** The observed association in U.S. studies likely reflects broader Western dietary patterns high in processed foods and refined carbohydrates, rather than a direct effect of eggs themselves.

○ **Conclusion:** The scientific evidence does not support the claim that
moderate egg consumption poses a significant diabetes risk for most
individuals. Observed associations are best explained by confounding
dietary and lifestyle factors, not by eggs per se. Randomized controlled
trials in healthy populations show only modest, clinically insignificant
changes in lipid profiles with moderate egg intake [114].

— The claim that "one egg per week increases diabetes risk by 76%" is
a mischaracterization of the evidence. Moderate egg consumption, as
part of a balanced diet, is not causally linked to increased diabetes
risk in the general population.

3.11   MSG (Monosodium Glutamate)

A common claim suggests that MSG (monosodium glutamate) and its related
compounds (disodium inosinate and disodium guanylate) are harmful food
additives that should be avoided at all costs. Proponents of this view argue
that these compounds "hijack your taste buds and make you crave to have
more," implying that they create addictive-like behaviors and compulsive
eating patterns. This narrative often appears in discussions about processed
foods, bouillon cubes, and restaurant meals, with claims that MSG causes
a range of adverse health effects and should be eliminated from the diet
entirely. The argument suggests that these flavor enhancers are inherently
dangerous and that any amount of consumption poses health risks.

This claim is often reinforced by references to "Chinese Restaurant
Syndrome," a term that emerged from anecdotal reports in the late 1960s,
and by the assertion that MSG is an "unnatural" chemical additive that
disrupts normal taste perception and eating behavior.

Key Point: MSG is a flavor enhancer that provides umami
(savory) taste and has been used safely in foods for over a
century. Understanding the actual evidence requires examining
controlled studies and regulatory assessments rather than relying
on anecdotal reports or outdated claims.

3.11.1   Scientific Analysis and Rebuttal

The claims about MSG being harmful and addictive do not hold up to
scientific scrutiny. Multiple comprehensive reviews and consensus statements

have concluded that MSG is considered safe for the vast majority of people at normal dietary intakes. Regulatory authorities worldwide, including the U.S. Food and Drug Administration (FDA), European Food Safety Authority (EFSA), and the Joint FAO/WHO Expert Committee on Food Additives (JECFA), have all reviewed the evidence and found MSG to be safe for use as a food additive [116–118].

There is no robust scientific evidence that MSG or its related compounds "hijack" taste buds or cause addiction or abnormal cravings in humans. While MSG does enhance the umami flavor and increases food palatability, this effect is similar to that of salt or sugar and does not equate to addiction or compulsive eating. Human studies consistently show that MSG can increase the palatability of foods, but the impact on food intake and satiety is inconsistent, with some studies even reporting enhanced satiety and reduced subsequent food intake [116, 118].

The term "Chinese Restaurant Syndrome" originated from a speculative 1968 letter in the New England Journal of Medicine and was not based on systematic research. Subsequent double-blind, placebo-controlled studies have not found consistent evidence that MSG causes the symptoms described, and the syndrome is not recognized as a formal medical diagnosis [119]. The focus on MSG in Chinese cuisine contributed to harmful stereotypes and racist narratives, as the same ingredient is widely used in Western processed foods without similar concern [116–118].

> **Key Point:** Controlled studies show that MSG does not cause addiction or harmful health effects at normal consumption levels. The scientific consensus from regulatory agencies worldwide supports MSG's safety for the general population.

3.11.2   Detailed Evidence: Safety and Regulations

Multiple lines of evidence support the safety of MSG and its related compounds when consumed at typical dietary levels.

Regulatory Safety Assessments: Comprehensive reviews by major food safety authorities consistently conclude that MSG is safe for the general population. The FDA classifies MSG as "generally recognized as safe" (GRAS), and similar assessments have been made by EFSA, Health Canada, and Food Standards Australia New Zealand (FSANZ). These agencies have set acceptable daily intake levels for MSG that are well above typical consumption levels. For example, in Europe, the mean intake of added MSG is 0.3–0.5 g/day, with high consumers reaching up to 1 g/day, and even in Asian countries, average intake is 1.2–1.7 g/day, all well below levels associated with adverse effects [116, 118].

Systematic Reviews and Toxicology Studies: Systematic reviews and peer-reviewed studies have found no significant health risks associated with MSG at typical consumption levels. The median lethal dose (LD50) of MSG in animal studies is significantly higher than that of table salt, indicating low acute toxicity. Adverse effects reported in animal studies occur at doses far higher than those encountered in human diets. A consensus meeting concluded that a maximum intake of 16,000 mg/kg body weight is regarded as safe, with no evidence of harm at typical dietary levels [116].

Genotoxicity and Experimental Risks: Some experimental studies, such as those reviewed by Syed Imam, have reported genotoxic effects of MSG at high doses or under specific laboratory conditions, including increased chromosomal aberrations and markers of genetic damage. However, these effects are observed at exposure levels far exceeding those encountered in normal human diets, and their relevance to typical dietary exposure remains debated. The broader literature and regulatory consensus do not consider these findings sufficient to alter the safety status of MSG for the general population [116, 120].

Double-Blind Controlled Studies: Double-blind, placebo-controlled studies have failed to confirm the existence of Chinese Restaurant Syndrome as a reproducible reaction to MSG. Some sensitive individuals may experience mild, short-term symptoms after consuming large amounts of MSG, but these effects are rare, not considered serious, and not consistently reproducible [116–118].

MSG and Brain Function: Concerns that dietary MSG could increase brain glutamate concentrations or disrupt brain function are not supported by scientific evidence. The blood-brain barrier is highly effective at restricting the passage of glutamate from the bloodstream into the brain. Even when dietary MSG increases blood glutamate slightly, brain glutamate levels remain stable, and no functional disruptions are observed. This has been confirmed in both animal and human studies [116, 121].

Sodium Reduction Benefits: Research demonstrates that Research demonstrates that MSG contains about two-thirds less sodium than table salt and can be used to replace a portion of sodium chloride in foods, allowing for a reduction in overall sodium content by up to 25–32.5% (and sometimes more) without compromising taste This makes MSG a valuable tool for reducing dietary sodium intake, which is beneficial for cardiovascular health [122, 123].

Related Compounds Safety: Disodium inosinate and disodium guanylate, the "cousins" of MSG mentioned in the claim, are also considered safe by food safety authorities at typical dietary levels. These compounds are widely used in the food industry and must be declared on food labels when added as ingredients [116].

> **Updated Key Point:** Decades of research and regulatory review show that MSG and its related compounds are safe for consumption and can even provide health benefits by enabling sodium reduction in foods. The scientific evidence directly contradicts claims about addiction, toxicity, or the need for complete avoidance.

3.11.3   Summary: Claim vs. Evidence

o **Claim:** MSG and its "cousins" (disodium inosinate, disodium guanylate) "hijack your taste buds," cause addictive cravings, and should be "avoided at all costs."

o **Rationale:** These compounds supposedly create abnormal taste responses and compulsive eating behaviors, making them inherently dangerous food additives.

o **Evidence:** Major food safety authorities worldwide classify MSG as **safe** for the general population, and controlled studies find **no evidence** of addiction or harmful effects at normal consumption levels.

o **Examples:**

– **Regulatory consensus:** FDA, EFSA, JECFA, and other agencies classify MSG as **"generally recognized as safe"** (GRAS) [116, 118].

– **Addiction claims:** Human studies show MSG increases palatability but find **no evidence** of addictive behaviors or abnormal cravings [117, 118].

– **Chinese Restaurant Syndrome:** Double-blind studies **failed to confirm** the syndrome as a reproducible reaction to MSG, and the narrative has been recognized as rooted in anecdote and cultural bias rather than science [116–118].

- **Health benefits:** MSG can **reduce sodium content** by 25–32.5% in foods without compromising taste, supporting cardiovascular health [122, 123].

- **Brain safety:** Dietary MSG does not increase brain glutamate concentrations or disrupt brain function due to the effectiveness of the blood-brain barrier [116, 121].

- **Genotoxicity:** Some experimental studies report genotoxic effects at high doses, but these are not relevant to normal dietary exposure and do not alter the regulatory consensus on MSG safety [116, 120].

○ **Proposed Mechanism:** MSG enhances umami taste naturally and safely, similar to how salt enhances saltiness or sugar enhances sweetness, without creating addiction or toxicity.

○ **Conclusion:** MSG and its related compounds are **safe and beneficial** flavor enhancers that can improve food taste while reducing sodium intake.

- The scientific evidence **directly contradicts** the fear-based claims. Rather than being avoided, MSG can be a valuable tool for creating flavorful, lower-sodium foods.

- The persistent myths about MSG largely stem from **anecdotal reports, cultural bias, and outdated information** rather than scientific evidence, demonstrating how misinformation can persist despite robust safety data.

3.12 Oats Consumption

A popular claim circulating online asserts that oats are "mostly starch" and therefore cause harmful spikes in blood sugar, especially in people with diabetes. Some go further, arguing that oats are inflammatory" and that switching from oats to eggs for breakfast leads to a reduction in all inflammation markers and a lower risk of heart disease. The implication is that oats are metabolically harmful, while eggs are inherently anti-inflammatory and protective for heart health.

3.12.1   Scientific Analysis and Rebuttal

The claim that oats are "inflammatory" and that switching to eggs for breakfast universally reduces inflammation and heart disease risk does not hold up under scientific scrutiny. Direct comparisons in controlled clinical trials show that eating oats or eggs for breakfast leads to similar outcomes in people with type 2 diabetes. There are no significant differences in weight, blood sugar, cholesterol, or C-reactive protein (CRP), i.e., the main blood marker doctors use to assess inflammation and heart disease risk. While one immune-related protein (tumor necrosis factor, or TNF) was found to be lower after the egg period, this does not mean oats are inflammatory, especially since the study did not measure TNF levels before the intervention. Larger reviews of the scientific literature show that oats can actually improve blood sugar control and cholesterol, and may even reduce inflammation in people with metabolic risk factors. There is no evidence that oats are harmful or inflammatory [124–126].

> **Key Point:** Direct comparisons between oats and eggs in people with type 2 diabetes show no difference in weight, blood sugar, cholesterol, or CRP (the main inflammation marker). Oats are not inflammatory; in fact, they improve blood sugar control and cholesterol, and may reduce inflammation in some at-risk groups.

3.12.2   Detailed Evidence: Trials and Analyses

A well-designed clinical trial in adults with type 2 diabetes directly compared the effects of eating an egg-based breakfast versus an oatmeal-based breakfast (with lactose-free milk) over two separate five-week periods. The researchers measured a range of health markers, including fasting blood sugar, long-term blood sugar control (measured by HbA1c), insulin, cholesterol, body weight, body fat, blood pressure, and C-reactive protein (CRP), which is the main blood marker doctors use to assess inflammation and heart disease risk. The results showed no significant differences between the two breakfasts for any of these measures. In other words, eating eggs or oatmeal for breakfast did not lead to changes in blood sugar, cholesterol, weight, or the main marker of inflammation. The only difference observed was that one immune-related protein, called tumor necrosis factor (TNF), was lower after the egg period, and another marker, interleukin-6 (IL-6), was borderline lower.

However, since the study did not measure TNF levels before the intervention, it is not possible to say that oatmeal increased inflammation—only that eggs may have reduced this particular marker slightly more during the study. Importantly, there were no differences in CRP, which is the most relevant marker for overall inflammation and heart disease risk. The authors concluded that eggs do not have negative effects on blood sugar or cholesterol in people with well-controlled type 2 diabetes, and while eggs may reduce some immune-related proteins compared to oatmeal, this effect is limited and not seen for the main inflammation marker (CRP) [124].

Looking more broadly at the scientific literature, a systematic review and meta-analysis of 23 randomized controlled trials examined whether eating oats affects inflammation-related markers in the body. The researchers found that, overall, oat intake did not significantly change levels of inflammation markers in the general population. However, when they looked specifically at people with existing health problems, such as high cholesterol or other metabolic issues, they found that oats significantly reduced CRP and another immune-related protein, IL-6. This suggests that oats may help lower inflammation in people who already have metabolic risk factors, but do not increase inflammation in healthy individuals. The authors concluded that while the overall evidence for oats reducing inflammation is modest, there is some benefit for people with metabolic disturbances [125].

Another comprehensive review and meta-analysis focused on people with type 2 diabetes. This analysis included 16 studies and found that eating oats significantly improved several important health markers. Specifically, oats lowered long-term blood sugar (HbA1c), fasting blood sugar, total cholesterol, and LDL cholesterol (the "bad" cholesterol) compared to control diets. Oatmeal also led to lower spikes in blood sugar and insulin after meals. There were no significant effects on HDL cholesterol, triglycerides, weight, or body mass index. The authors concluded that oats have clear benefits for blood sugar control and cholesterol in people with type 2 diabetes [126].

Thus, the best available evidence from clinical trials and meta-analyses shows that oats do not increase inflammation and, in some cases, may help reduce it, especially in people with existing metabolic risk factors. Oats also consistently improve blood sugar control and cholesterol in people with type 2 diabetes. Eggs may reduce certain immune-related proteins compared to oatmeal, but do not affect the main marker of inflammation or other key health measures. Both oats and eggs can be part of a healthy diet, and there is no evidence that oats are harmful or inflammatory.

Key Point: Oats improve blood sugar and cholesterol in type 2 diabetes, and may reduce inflammation in people with metabolic risk. Eggs may reduce some immune-related proteins compared to oatmeal, but do not affect the main inflammation marker (CRP) or other risk markers.

3.12.3 Summary: Claim vs. Evidence

o **Claim:** Oats are inflammatory" and cause blood sugar spikes; switching from oats to eggs reduces all inflammation markers and heart disease risk.

o **Rationale:** Oats are mostly starch, leading to blood sugar spikes and inflammation, while eggs are anti-inflammatory and protective.

o **Evidence:**

  – **Randomized trial:** In people with type 2 diabetes, eggs and oatmeal had **no difference** in weight, blood sugar, cholesterol, or CRP (the main inflammation marker). Only one immune-related protein (TNF) was lower after eggs, but no baseline was measured, so oats cannot be called inflammatory [124].

  – **Meta-analyses:** Oats **improve blood sugar control and cholesterol** in type 2 diabetes, and may reduce CRP and IL-6 in people with metabolic risk, but do not increase inflammation [125, 126].

  – **No evidence** that oats increase heart disease risk or overall inflammation.

o **Proposed Mechanism:** Oats are rich in soluble fiber (beta-glucan), which slows sugar absorption and lowers cholesterol. Eggs are a source of high-quality protein and micronutrients, and may reduce some immune-related proteins, but do not outperform oats in overall metabolic health.

o **Conclusion:** The scientific evidence **contradicts** the claim that oats are inflammatory or harmful for people with diabetes. Both oats and eggs can be part of a healthy diet; oats provide clear benefits for blood sugar and cholesterol, and neither food increases overall inflammation or heart disease risk.

  – Claims that oats are inflammatory" or that eggs universally reduce heart disease risk are not supported by controlled studies. Dietary recommendations should be based on the totality of evidence, not isolated or exaggerated claims.

3.13 Red Meat and Cancer Risk

A common argument in popular health discussions is that there is no convincing evidence linking red meat consumption to cancer risk. Proponents of this view often suggest that red meat may even be protective against cancer, citing its unique nutrient profile. They also point out that there are no randomized controlled trials in humans showing that eating more red meat increases inflammation or cancer risk, and therefore conclude that concerns about red meat are unfounded.

However, this debate reveals a striking paradox in how scientific evidence is interpreted (and often misinterpreted) online. If the scientific evidence were reversed, and robust studies consistently showed that red meat was harmless or even beneficial, it is likely that many of the same voices currently defending red meat would instead be loudly warning of its dangers. We see a similar phenomenon with seed oils: despite a lack of evidence that seed oils are harmful, large segments of the internet claim they are toxic and responsible for a host of modern diseases. The arguments often hinge not on the actual weight of scientific evidence, but on the desire to take a contrarian stance or to align with a particular dietary ideology.

This tendency is not unique to red meat or seed oils. It reflects a broader pattern in nutrition debates, where the interpretation of evidence is often shaped more by cultural, social, or ideological factors than by the actual data. When the evidence points in one direction, online communities frequently rally around the opposite position, dismissing decades of research in favor of anecdote or speculation.

3.13.1 Scientific Analysis and Rebuttal

This reasoning does not align with the way cancer research is conducted or with the current scientific evidence. Logically, randomized controlled trials directly assigning people to eat high or low amounts of red meat for decades, because cancers typically take many years to develop, just to observe cancer outcomes are not feasible. such trials would be impractical and unethical. This is not unique to red meat. The same logic would dismiss the well-established link between smoking and lung cancer, for which no randomized trials exist. Instead, scientists rely on large, long-term prospective studies that follow people over many years, as well as meta-analyses that combine data from many such studies. These are considered the gold standard for evaluating long-term dietary exposures and cancer risk.

Multiple comprehensive meta-analyses and large cohort studies have consistently found that higher consumption of red and processed meats is associated with a significantly increased risk of colorectal cancer, as well as increased risks for colon and rectal cancers specifically. These associations remain even after accounting for other factors such as smoking, alcohol intake, physical activity, and other dietary habits.

> **Key Point:** The absence of randomized controlled trials for diet and cancer does not mean there is no evidence. Large, long-term studies consistently show that eating more red and processed meat increases the risk of colorectal and other cancers.

3.13.2 Detailed Evidence: Analyses and Studies

Comprehensive Meta-Analysis (2025): A 2025 meta-analysis that included 60 prospective studies found that people who ate the most red meat had a 22% higher risk of colon cancer, a 15% higher risk of colorectal cancer overall, and a 22% higher risk of rectal cancer compared to those who ate the least. For processed meat, the increased risks were similar: 13% higher for colon cancer, 21% higher for colorectal cancer, and 17% higher for rectal cancer. When looking at total meat consumption, the risks were also elevated by about 19% for these cancers. These results were consistent across different populations and study designs, and the analysis found no evidence that the results were skewed by publication bias. The authors concluded that these findings strongly support current dietary recommendations to limit red and processed meat intake as part of cancer prevention strategies [127].

Systematic Review and Meta-Analysis (2021): A 2021 review of 148 prospective studies found that high red meat intake was associated with a 9% higher risk of breast cancer, a 25% higher risk of endometrial cancer, a 10% higher risk of colorectal cancer, a 17% higher risk of colon cancer, a 22% higher risk of rectal cancer, a 26% higher risk of lung cancer, and a 22% higher risk of liver cancer. High processed meat intake was also linked to higher risks of breast, colorectal, colon, rectal, and lung cancers. When red and processed meat were combined, the risk of colorectal cancer was 17% higher for those eating the most compared to those eating the least [128].

UK Biobank Prospective Cohort Study: A large study of nearly half a million people in the UK found that eating 70 grams more red and processed meat per day (about the size of a small steak or a couple of slices of bacon) was associated with a 32% higher risk of colorectal cancer and a 40% higher risk of colon cancer. These results remained even after adjusting for other lifestyle and dietary factors. The study found little evidence that meat intake was associated with risk of other cancers, supporting the specificity of the association with colorectal cancer [129].

3.13.3 Biological Mechanisms

The link between red and processed meat and cancer is supported by well-established biological mechanisms. Red and processed meats contain heme iron, which can promote the formation of carcinogenic compounds in the gut. Cooking and processing meats at high temperatures generate chemicals such as heterocyclic amines and polycyclic aromatic hydrocarbons, which can damage DNA. Processed meats often contain nitrates and nitrites, which can also form carcinogens. These mechanisms are supported by both experimental and epidemiological evidence [127].

> **Key Point:** Eating more red and processed meat increases the risk of colorectal, colon, and rectal cancers by about 13–22%, according to the largest and most comprehensive studies to date. The evidence is strongest for colorectal cancer, and the biological mechanisms are well understood.

3.13.4 Summary: Claim vs. Evidence

○ **Claim:** There is no convincing evidence that red meat causes cancer, and it may even be protective. The lack of randomized controlled trials means causation cannot be established.

○ **Rationale:** Red meat contains beneficial nutrients, and without controlled trials showing direct causation, observational studies are seen as insufficient to prove harm.

○ **Evidence:** Multiple large-scale meta-analyses and cohort studies show that people who eat the most red and processed meat have a 13–22%

higher risk of colorectal, colon, and rectal cancers compared to those who eat the least. These findings are consistent, robust, and biologically plausible [127–129].

○ **Examples:**

– In the 2025 meta-analysis, high red meat intake was linked to a 22% higher risk of colon cancer, a 15% higher risk of colorectal cancer, and a 22% higher risk of rectal cancer.

– The 2021 review found a 10–22% higher risk of colorectal, colon, and rectal cancers for those eating the most red meat.

– The UK Biobank study found that eating 70 grams more red and processed meat per day was associated with a 32% higher risk of colorectal cancer and a 40% higher risk of colon cancer.

○ **Proposed Mechanism:** Red and processed meats increase cancer risk through the formation of carcinogenic compounds during cooking and digestion, oxidative stress from heme iron, and inflammatory responses.

○ **Conclusion:** The scientific evidence strongly supports an association between higher red and processed meat consumption and increased cancer risk, especially for colorectal cancer. The absence of randomized controlled trials does not negate the substantial evidence from prospective studies, which consistently show increased risk.

– The scientific evidence directly contradicts the claim of no association. Dietary guidelines recommending limits on red and processed meat intake are well supported by the best available research.

3.14   Sugar Consumption and Cancer Risk

A persistent claim in popular health discourse is that "sugar feeds cancer" and that reducing sugar intake can prevent, treat, or even cure cancer. This narrative is often accompanied by warnings against the use of calorie-dense, sweetened drinks for people with cancer, with the implication that such beverages accelerate tumor growth or worsen outcomes. The argument is typically based on the observation that cancer cells are highly metabolically active and require glucose for energy, leading to the belief that dietary sugar restriction will "starve" cancer cells while sparing healthy tissue.

However, this claim is a classic example of oversimplification and misinterpretation of metabolic science. While it is true that cancer cells consume more glucose than most normal cells, all cells in the body require glucose to function, and there is no evidence that dietary sugar restriction can selectively deprive cancer cells of energy without also harming healthy tissues. Moreover, the relationship between sugar intake, obesity, and cancer risk is complex and mediated by multiple factors, including overall calorie balance, body weight, and metabolic health.

> **Key Point:** Cancer cells do require more glucose than normal cells, but there is no evidence that reducing sugar intake treats or cures cancer. Most studies find no direct link between sugar consumption and cancer risk, and the observed associations are largely explained by excess calorie intake and weight gain, not sugar itself.

3.14.1   Scientific Analysis and Rebuttal

The claim that dietary sugar restriction can treat or cure cancer is not supported by scientific evidence. A comprehensive systematic review of 37 prospective cohort studies found that the majority reported no association between total sugar, sucrose, or fructose intake and cancer risk [130]. Where associations were observed, they were generally weak and appeared to be mediated by excess adiposity and weight gain, rather than by sugar per se. The review also found that higher intake of added sugars and sugary beverages was sometimes associated with increased cancer risk, but these findings were inconsistent and often confounded by other dietary and lifestyle factors.

Large umbrella reviews and meta-analyses confirm these findings. For example, a 2023 umbrella review found that high dietary sugar consumption is generally more harmful than beneficial for health, especially in relation to cardiometabolic disease, but the evidence linking sugar intake to cancer risk remains limited and of low quality [131]. The review concluded that while there is some suggestive evidence for a higher risk of certain cancers (such as pancreatic cancer with high fructose intake), most associations between sugar and cancer are weak, inconsistent, and likely confounded by obesity and other risk factors.

A large prospective cohort study from France (NutriNet-Santé, $n >$ $100,000$) found that higher consumption of sugary drinks (including 100%

fruit juice) was associated with a modestly increased risk of overall cancer and breast cancer, but not with prostate or colorectal cancer [132]. Importantly, the study found no association between artificially sweetened beverages and cancer risk. The authors noted that the observed associations were likely driven by the sugar content of the drinks and by their contribution to excess calorie intake and weight gain, rather than by a direct effect of sugar on cancer cells.

Meta-analyses of observational studies on sweet beverage consumption and cancer risk have found similar results. A 2021 systematic review and meta-analysis reported a small but statistically significant association between high sugar-sweetened beverage (SSB) intake and increased risk of breast and prostate cancer, but not for most other cancer types [133]. The authors emphasized that the associations were modest, and that the evidence for a direct causal link between sugar and cancer remains weak.

> **Key Point:** Most studies find no direct link between sugar intake and cancer risk. Where associations exist, they are small and likely explained by excess calorie intake and weight gain, not by sugar itself. There is no evidence that reducing sugar intake treats or cures cancer.

3.14.2 Detailed Evidence: Reviews, Analyses, and Studies

Systematic Review of Prospective Studies: Makarem et al. (2018) systematically reviewed 37 prospective cohort studies on dietary sugars and cancer risk. The majority of studies on total sugar, sucrose, and fructose intake reported null associations with cancer risk. Of the few studies that found increased risk, the associations were generally limited to added sugars and sugary beverages, and were often explained by increased body weight and adiposity rather than sugar itself. The review concluded that "most studies were indicative of a null association, but suggestive detrimental associations were reported for added sugars and sugary beverages" [130].

Umbrella Review of Meta-Analyses: Huang et al. (2023) conducted an umbrella review of 73 meta-analyses, including 83 health outcomes. They found that high sugar consumption was associated with increased risk of several metabolic and cardiovascular diseases, but the evidence for cancer was limited and of low quality. For example, each 25 g/day increment of fructose was associated with a 22% higher risk of pancreatic cancer (class III evidence), but most other associations between sugar and cancer were weak or non-significant. The review concluded: "Evidence of the association between dietary sugar consumption and cancer remains limited but warrants further research" [131].

Large Prospective Cohort Study (NutriNet-Santé): Chazelas et al. (2019) followed over 100,000 adults for a median of 5.1 years and found that each 100 mL/day increase in sugary drink consumption was associated with an 18% higher risk of overall cancer and a 22% higher risk of breast cancer. However, the authors noted that these associations were likely mediated by increased calorie intake and weight gain, and that artificially sweetened beverages were not associated with cancer risk. The study concluded: "These results need replication in other large scale prospective studies. They suggest that sugary drinks, which are widely consumed in Western countries, might represent a modifiable risk factor for cancer prevention" [132].

Meta-Analysis of Sweet Beverage Consumption: Llaha et al. (2021) analyzed 27 studies and found that people who consumed the most sugar-sweetened beverages had about a 14% higher risk of breast cancer and an 18% higher risk of prostate cancer compared to those who consumed the least. The authors emphasized that these associations were modest (i.e., not a doubling or tripling of risk) and that the evidence for a direct causal link between sugar and cancer remains weak [133].

Cachexia and Cancer Nutrition: A critical and often overlooked aspect of cancer care is the management of cachexia, a syndrome characterized by severe muscle and fat loss that affects up to 80% of people with advanced cancer [134]. Cachexia is a major cause of morbidity and mortality in cancer patients, and one of its most important causes is a reduction in calorie intake. Creating fear around calorie-dense, easy-to-consume drinks in cancer patients is dangerous, as inadequate nutrition can worsen cachexia and increase the risk of death. In fact, around a third of people with cancer die due to cachexia, often from heart or lung failure. Current guidelines emphasize the importance of maintaining adequate calorie and protein intake in cancer patients, and there is no evidence that restricting sugar or carbohydrate intake improves outcomes in this population [134].

> **Key Point:** For people with cancer, especially those at risk of cachexia, maintaining adequate calorie and protein intake is

critical. Fear-based avoidance of calorie-dense drinks can be dangerous and may worsen outcomes.

3.14.3   Summary: Claim vs. Evidence

o **Claim:** "Sugar feeds cancer," and reducing sugar intake can treat or cure cancer. Cancer patients should avoid calorie-dense, sweetened drinks.

o **Rationale:** Cancer cells require more glucose than normal cells, so restricting sugar will "starve" the tumor.

o **Evidence:**

   - **Systematic reviews and meta-analyses:** The majority of studies find no association between total sugar, sucrose, or fructose intake and cancer risk [130, 131].

   - **Where associations exist:** They are small, inconsistent, and largely explained by excess calorie intake and weight gain, not by sugar itself [132, 133].

   - **Cancer cachexia:** Up to 80% of people with advanced cancer experience cachexia, and inadequate calorie intake is a major cause. Fear-based avoidance of calorie-dense drinks can worsen cachexia and increase mortality [134].

o **Proposed Mechanism:** All cells, including healthy ones, require glucose. There is no way to selectively "starve" cancer cells of glucose through diet without also harming normal tissues. The observed links between sugar and cancer are mediated by obesity and excess calorie intake, not by sugar itself.

o **Conclusion:** There is no evidence that reducing sugar intake treats or cures cancer. Most studies find no direct link between sugar and cancer risk, and the observed associations are explained by excess calorie intake and weight gain. For cancer patients, maintaining adequate nutrition is critical, and fear-based dietary restrictions can be dangerous.

   - The scientific evidence directly contradicts the claim that sugar restriction treats cancer. Dietary advice for cancer patients should focus on adequate calorie and protein intake, not on unnecessary sugar avoidance.

3.15   Plant-Based Meats

A frequently repeated claim in online health circles, sometimes echoed by medical professionals, is that plant-based meats are "ultra-processed," "full of synthetic ingredients and artificial flavors," and "loaded with pro-inflammatory seed oils." The argument often concludes that these products are "nothing like real food" and should be avoided for optimal health. Proponents of this view typically invoke the "naturalistic fallacy," i.e., the idea that only foods close to their natural state are healthy, while warning that the ingredients and processing methods used in plant-based meats make them inherently unhealthy or even dangerous.

3.15.1   Scientific Analysis and Rebuttal

This claim does not withstand scientific scrutiny. The focus on whether a food is "natural" or "synthetic" is a superficial argument that ignores the actual nutritional profile and health effects of the food in question. Many natural substances (such as certain mushrooms, nightshade plants, or heavy metals) are toxic, while many processed foods can be health-promoting depending on their composition and context. The relevant question is not whether a food is "natural," but how it affects health markers and disease risk in controlled studies.

Recent high-quality evidence directly addresses the health effects of plant-based meat alternatives (PBMAs) compared to animal meats. A 2025 systematic review and meta-analysis of seven randomized controlled trials found that substituting PBMAs for meat in the diet led to significant reductions in LDL cholesterol, total cholesterol, and body weight, with no adverse effects on other cardiometabolic markers. These findings were consistent across studies using both soy/legume-based and mycoprotein-based alternatives, and were robust in sensitivity analyses [135].

Further, a 2024 review in the Canadian Journal of Cardiology concluded that PBMAs are generally lower in saturated fat and higher in polyunsaturated fats (PUFAs) and dietary fiber compared to meat. Some dietary trials replacing meat with PBMAs reported improvements in cardiovascular risk factors, including reductions in total cholesterol, LDL cholesterol, apolipoprotein B-100, and body weight. Importantly, there is no evidence that the processing or sodium content of PBMAs negates these potential cardiovascular benefits [136].

The claim that "seed oils" (i.e., oils high in linoleic acid, an omega-6 PUFA) used in PBMAs are "pro-inflammatory" is also not supported by the scientific literature. A 2017 meta-analysis of 30 randomized controlled trials found that increasing dietary linoleic acid intake did not significantly affect blood concentrations of inflammatory markers, including C-reactive protein (CRP), interleukin-6, tumor necrosis factor, or adhesion molecules. Subgroup analyses suggested that only extremely high increases in linoleic acid might slightly raise CRP, but this was not observed in the majority of studies [137]. Another controlled trial comparing high-PUFA (seed oil–rich) diets to saturated fat–rich diets found that PUFAs reduced liver fat, improved cholesterol profiles, and did not increase markers of inflammation or oxidative stress. In fact, some inflammatory markers were lower on the PUFA-rich diet [138].

Key Points:

o The "natural vs. synthetic" argument is a logical fallacy; health effects depend on nutritional profile and clinical outcomes, not on whether a food is "natural."

o Plant-based meats are generally lower in saturated fat and higher in fiber and PUFAs than animal meats.

o Replacing meat with plant-based meats in randomized controlled trials leads to lower LDL cholesterol, total cholesterol, and body weight, with no adverse effects on other blood markers.

o Seed oils (rich in linoleic acid) do not increase inflammation in humans, according to multiple meta-analyses of randomized controlled trials.

3.15.2   Detailed Evidence: Summary of Studies

Systematic Review and Meta-Analysis (2025): A meta-analysis of seven randomized controlled trials (369 adults, 1–8 weeks duration) found that replacing meat with PBMAs reduced LDL cholesterol by 0.25 mmol/L, total cholesterol by 0.29 mmol/L, and body weight by 0.72 kg. No significant changes were observed in HDL cholesterol, triglycerides, blood pressure, or fasting glucose. Sensitivity analyses confirmed these findings for both soy/legume-based and mycoprotein-based PBMAs. The authors concluded that PBMAs may facilitate the transition to a plant-based diet and are likely cardioprotective, but longer-term studies are needed [135].

Review of Nutritional Profiles and Cardiovascular Outcomes (2024): A review in the Canadian Journal of Cardiology found that PBMAs are typically lower in saturated fat and higher in polyunsaturated fat and fiber than meat. Trials replacing meat with PBMAs reported improvements in cardiovascular risk factors, including cholesterol and body weight. No evidence suggests that the processing or sodium content of PBMAs negates these benefits. The review concluded that replacing meat with PBMAs may be cardioprotective, but called for longer-term studies to confirm effects on cardiovascular events [136].

Meta-Analysis of Linoleic Acid and Inflammation (2017): A meta-analysis of 30 randomized controlled trials (1,377 subjects) found no significant effect of higher linoleic acid intake on inflammatory markers (CRP, interleukin-6, TNF, adiponectin, or adhesion molecules). Only in subgroups with extremely high increases in linoleic acid was a slight increase in CRP observed, but this was not the norm. The authors concluded that increasing dietary linoleic acid does not have a significant effect on blood concentrations of inflammatory markers [137].

RCT: PUFAs vs. Saturated Fats (2012): A 10-week RCT in abdominally obese adults compared a high-PUFA (seed oil–rich) diet to a saturated fat–rich diet. The PUFA diet reduced liver fat, improved cholesterol profiles, and did not increase markers of inflammation or oxidative stress. In fact, some inflammatory markers (IL-1 receptor antagonist and TNF receptor-2) were lower on the PUFA diet. The authors concluded that high n-6 PUFA intake does not cause inflammation or oxidative stress and may be beneficial for metabolic health [138].

> **Key Point:** Controlled trials and meta-analyses show that plant-based meats, when used to replace animal meats, improve cholesterol and body weight without increasing inflammation. Seed oils used in these products do not cause inflammation in humans.

3.15.3   Summary: Claim vs. Evidence

- **Claim:** Plant-based meats are "ultra-processed," "full of synthetic ingredients and seed oils," and are "pro-inflammatory" and unhealthy.

- **Rationale:** The argument relies on the naturalistic fallacy and the belief that processing or seed oils inherently cause harm.

o **Evidence:**

- **Systematic review/meta-analysis:** Replacing meat with plant-based meats in randomized controlled trials lowers LDL cholesterol, total cholesterol, and body weight, with no adverse effects on other blood markers [135].

- **Nutritional profile review:** PBMAs are lower in saturated fat and higher in fiber and PUFAs than meat; no evidence that processing or sodium content negates benefits [136].

- **Inflammation research:** Meta-analyses of randomized controlled trials show that linoleic acid (from seed oils) does not increase inflammatory markers in humans [137, 138].

o **Proposed Mechanism:** PBMAs improve cardiovascular risk factors by reducing saturated fat and increasing fiber and PUFAs; seed oils do not cause inflammation.

o **Conclusion:** The scientific evidence directly contradicts the claim that plant-based meats are inherently unhealthy or pro-inflammatory. When used to replace meat, they improve cholesterol and body weight, and seed oils do not increase inflammation. Health effects should be judged by clinical outcomes, not by whether a food is "natural" or "synthetic."

- The "natural vs. synthetic" argument is a distraction from the real question: what does the evidence show about health outcomes? In the case of plant-based meats, the evidence supports their safety and potential benefits when used as a replacement for animal meats.

3.16   Insulin Resistance and Weight Loss

A persistent claim in popular health discussions is that insulin resistance is the primary reason people struggle to lose weight, even when they are dieting and exercising. Proponents argue that high insulin levels "block" fat burning, making it "impossible" to lose weight unless insulin is lowered. Some go further, asserting that calories-in-versus-calories-out is "meaningless" in the context of insulin resistance, and that only by reducing insulin (typically via low-carbohydrate or ketogenic diets) can fat loss occur. This narrative is often accompanied by the assertion that insulin resistance lowers metabolism, trapping individuals in a cycle of weight gain and metabolic dysfunction.

Key Point: The claim is that high insulin or insulin resistance prevents fat loss and lowers metabolism, so that calorie restriction is ineffective unless insulin is reduced.

3.16.1 Scientific Analysis and Rebuttal

This claim does not withstand scrutiny when examined in light of controlled metabolic studies and clinical trials. Multiple lines of evidence show that:

○ High insulin levels or insulin resistance do **not** prevent weight loss when calories are restricted.

○ In fact, insulin resistance is associated with a **higher** resting metabolic rate (RMR), not a lower one.

○ Isocaloric (calorie-matched) low-carbohydrate and low-fat diets produce similar weight and fat loss when protein and calories are held constant, regardless of insulin levels or insulin resistance.

Resting Metabolic Rate and Insulin Resistance: Contrary to the claim that insulin resistance lowers metabolism, several studies demonstrate the opposite. Obese individuals with type 2 diabetes (a state of insulin resistance) have a significantly higher RMR than obese non-diabetic controls. For example, a case-control study in Sudan found that obese diabetic patients had a higher RMR (1481kcal/day) than obese non-diabetics (1362kcal/day), with the highest RMRs observed in those with poor glycemic control [139]. This finding is consistent with other studies in both Western and Asian populations, and is supported by a comprehensive review which found that uncontrolled diabetes is associated with a 7.7% higher basal energy expenditure (BEE, i.e., the energy expended to maintain minimal metabolic activities during a non-active period) than healthy controls [140]. The likely mechanisms include increased gluconeogenesis, protein turnover, and sympathetic nervous system activity in insulin-resistant states.

A large cohort study of 782 adults found a strong association between increasing insulin resistance (measured by HOMA-IR) and higher RMR, even after adjusting for age, sex, and fat-free mass [141]. This relationship held true across both sexes and all BMI categories. The authors concluded that insulin resistance is associated with higher, not lower, metabolic rate.

Isocaloric Diets and Weight Loss: The claim that calories in versus calories out means "nothing" in the context of insulin resistance is directly contradicted by randomized controlled trials. In tightly controlled metabolic ward studies, participants assigned to high-carbohydrate or low-carbohydrate diets with equal calories and protein lost similar amounts of weight and body fat. For example, Hall et al. [142] found that switching overweight men from a high-carbohydrate to an isocaloric ketogenic diet (with protein held constant) did **not** increase body fat loss, despite a small, transient increase in energy expenditure. Both diets produced similar overall fat loss, and the difference in energy expenditure was minor (about 50–100 kcal/day), far less than the 300–600 kcal/day predicted by the "carbohydrate-insulin" model.

Large free-living trials confirm these findings. In the DIETFITS trial, over 600 overweight adults were randomized to a healthy low-fat or healthy low-carbohydrate diet for 12 months. There was no significant difference in weight loss between groups (−5.3 kg vs. −6.0 kg), and neither baseline insulin secretion nor genotype predicted which diet worked better for whom [143]. A pilot trial with detailed insulin resistance phenotyping also found no significant interaction between diet type and insulin resistance status for weight loss [144]. Both insulin-resistant and insulin-sensitive participants lost substantial weight on both diets, provided they reduced calories and improved diet quality.

Low-Carbohydrate Diets in Diabetes: In people with type 2 diabetes, isocaloric low-carbohydrate diets can improve glycemic control and reduce insulin levels, but do not produce greater fat loss than isocaloric high-carbohydrate diets. Some studies suggest that low-carbohydrate diets may preferentially reduce visceral fat and improve HDL cholesterol, but total weight and fat loss are similar when calories are matched [145].

> **Key Point:** Insulin resistance is associated with a higher, not lower, resting metabolic rate. Weight loss occurs when calories are reduced, regardless of insulin levels or resistance. Isocaloric low-carb and low-fat diets produce similar fat loss.

3.16.2   Detailed Evidence: Summary of Studies

Resting Metabolic Rate: Obese diabetic patients have higher RMR than obese non-diabetics (1481 vs. 1362kcal/day, $p = 0.027$), especially with poor glycemic control [139]. Severely obese diabetics also have higher RMR than non-diabetics, even after adjusting for age, sex, weight, and height [146]. A review found that uncontrolled diabetes increases BEE by 7.7% [140].

A large cohort study found a strong positive association between insulin resistance and RMR [141].

Isocaloric Diets: In a metabolic ward study, switching from a high-carb to a ketogenic diet (with calories and protein held constant) did not increase body fat loss, and only slightly increased energy expenditure (~50–100kcal/day) [142]. Body fat loss actually slowed transiently on the ketogenic diet.

Free-Living Trials: In the DIETFITS trial ($n = 609$), there was no significant difference in 12-month weight loss between low-fat and low-carb diets, and neither genotype nor insulin secretion predicted success [143]. A pilot trial with insulin resistance phenotyping found no significant diet × insulin resistance interaction for weight loss [144].

Low-Carb Diets in Diabetes: In obese type 2 diabetics, isocaloric low-carb diets may reduce visceral fat and improve HDL-C more than high-carb diets, but total weight and fat loss are similar [145].

3.16.3   Summary: Claim vs. Evidence

- **Claim:** Insulin resistance prevents fat loss and lowers metabolism; calorie restriction is ineffective unless insulin is reduced.

- **Rationale:** High insulin "blocks" fat burning, so only low-carb diets work for weight loss in insulin-resistant people.

- **Evidence:**

  - Insulin resistance is associated with **higher** resting metabolic rate, not lower [139–141, 146].

  - Isocaloric low-carb and low-fat diets produce similar weight and fat loss when protein and calories are matched [142–144].

  - Large trials show no significant diet × insulin resistance interaction for weight loss [143, 144].

  - In type 2 diabetes, isocaloric low-carb diets may reduce visceral fat, but do not increase total fat loss compared to high-carb diets [145].

- **Proposed Mechanism:** Insulin resistance increases gluconeogenesis, protein turnover, and sympathetic activity, raising energy expenditure.

- **Conclusion:** The scientific evidence directly contradicts the claim. Insulin resistance does not "block" fat loss or lower metabolism. Weight loss occurs when calories are reduced, regardless of insulin levels or resistance. Both low-carb and low-fat diets can be effective if calories and protein are controlled.

> **Updated Key Point:** Insulin resistance is associated with higher, not lower, metabolism. In controlled studies, isocaloric low-carb and low-fat diets produce similar fat loss when calories and protein are matched, suggesting that insulin levels alone do not determine weight loss outcomes. However, weight loss is influenced by multiple complex factors including food matrix effects, individual metabolic responses, genetics, hormones, and adherence patterns.

3.17   Chia Seeds and Inflammation

Some sources claim that chia seeds, despite being rich in omega-3 fatty acids and fiber, may actually increase inflammation in humans, even as they raise blood omega-3 levels. This warning suggests that chia seeds could worsen inflammatory markers and should be consumed with caution.

3.17.1   Scientific Analysis and Rebuttal

This claim does not hold up when examined against the best available scientific evidence. Recent systematic reviews and meta-analyses of randomized controlled trials in humans have directly tested the effects of chia seed supplementation on key markers of inflammation, such as C-reactive protein (CRP), interleukin-6 (IL-6), and tumor necrosis factor-alpha (TNF-α). These studies consistently show that chia seed supplementation leads to a reduction in CRP, a major marker of systemic inflammation, and does not increase IL-6 or TNF-α. In fact, no controlled trial has found evidence that chia seeds increase inflammation in humans.

> **Key Point:** The most rigorous human studies show that chia seed supplementation reduces CRP, a key marker of inflammation, and does not increase IL-6 or TNF-α. There is no evidence from controlled trials that chia seeds increase inflammation in humans.

3.17.2 Detailed Evidence: Analyses and Trials

A recent systematic review and meta-analysis that focused exclusively on randomized controlled trials in adults found that chia seed supplementation led to a clear reduction in CRP, especially in people with type 2 diabetes, with longer intervention durations, and at higher doses. The effect was not seen for IL-6 or TNF-α, which remained unchanged with chia supplementation. The studies included in this analysis were conducted in both overweight individuals and those with type 2 diabetes, and the results were consistent across different populations and study designs. The overall quality of evidence was rated as moderate for CRP and TNF-α, and low for IL-6, mainly due to the small number of studies and some methodological concerns, but the direction of effect was consistent across trials [147].

Another comprehensive meta-analysis, which included a larger number of randomized controlled trials in overweight subjects, confirmed these findings. Chia seed supplementation was associated with a reduction in CRP, but had no significant effect on IL-6 or TNF-α. In addition to its anti-inflammatory effect, chia seed supplementation also led to reductions in waist circumference and systolic blood pressure, while having no significant impact on cholesterol, blood sugar, or body weight. These benefits were observed regardless of the form in which chia was consumed (whole, ground, or as flour), and the results were robust across different study designs [148].

Both meta-analyses highlight that the anti-inflammatory effect of chia seeds is specific to CRP, which is an acute-phase protein and a sensitive marker of systemic inflammation and cardiovascular risk. The lack of effect on IL-6 and TNF-α, which are cytokines involved in chronic inflammation, suggests that chia's benefits may be most relevant for reducing low-grade systemic inflammation, rather than altering chronic cytokine profiles.

Mechanistically, chia seeds are rich in alpha-linolenic acid (ALA, an omega-3 fatty acid), dietary fiber, and polyphenols, all of which have been shown to modulate inflammatory pathways. The reduction in CRP may be mediated by improvements in insulin sensitivity, antioxidant status, and modulation of inflammatory signaling. Importantly, no controlled trial has reported an increase in any inflammatory marker with chia supplementation.

> **Key Point:** Chia seed supplementation reduces CRP, a key marker of inflammation, and does not increase IL-6 or TNF-α. These effects are most pronounced with higher doses and longer interventions, especially in people with metabolic risk factors.

3.17.3   Summary: Claim vs. Evidence

o **Claim:** Chia seeds increase inflammation in humans, despite raising omega-3 levels.

o **Rationale:** Some anecdotal reports and misinterpretations of isolated studies suggest that chia seeds may worsen inflammatory markers.

o **Evidence:** Two recent meta-analyses of randomized controlled trials show that chia seed supplementation reduces CRP, a key marker of inflammation, while having no effect on other inflammatory markers such IL-6 or TNF-α. No controlled trial has found an increase in inflammatory markers with chia seeds [147, 148].

    – **CRP:** Reduced after chia supplementation, especially in people with type 2 diabetes and with higher doses or longer duration.

    – **IL-6, TNF-α:** No significant change.

    – **Other benefits:** Reductions in waist circumference and systolic blood pressure.

o **Proposed Mechanism:** Chia seeds provide ALA, fiber, and polyphenols, which reduce systemic inflammation via improved insulin sensitivity, antioxidant effects, and modulation of inflammatory signaling pathways.

o **Conclusion:** The scientific evidence directly contradicts the claim that chia seeds increase inflammation. On the contrary, chia seeds may reduce systemic inflammation (as measured by CRP) and improve other cardiometabolic risk factors. No controlled trial has found evidence of increased inflammation with chia supplementation.

    – Chia seeds are a safe and potentially beneficial addition to a balanced diet, especially for individuals with metabolic risk factors.

    Updated Key Point: Chia seed supplementation reduces CRP and does not increase other inflammatory markers. There is no evidence from controlled human trials that chia seeds increase inflammation; in fact, they may provide modest anti-inflammatory and cardiometabolic benefits.

3.18 Can Alcohol Be Beneficial?

A widely held belief in popular health discussions is that drinking one or two glasses of wine each night is essentially harmless, or may even be good for you, especially for your heart. This narrative is so pervasive that many people assume moderate alcohol consumption is a healthy lifestyle choice, often citing studies that seem to show moderate drinkers live longer or have lower rates of heart disease than those who abstain. The supposed benefits are usually attributed to observational studies that find moderate drinkers have better health outcomes than non-drinkers, leading to the conclusion that alcohol, in small amounts, is protective.

> **Key Point:** The claim that moderate alcohol consumption is harmless or beneficial is based on flawed observational studies. When these biases are corrected, the evidence shows that even low levels of alcohol are linked to increased health risks.

3.18.1 Scientific Analysis and Rebuttal

The idea that moderate drinking is safe or good for you does not hold up to rigorous scientific scrutiny. A major flaw in many studies is the "drinker misclassification error," sometimes called the "sick quitter" problem. In a large review of over a hundred studies, most included people who used to drink (and often quit due to health problems) in the group labeled as "abstainers." This means the non-drinking group actually included many people who had health issues, making moderate drinkers look healthier by comparison. When studies properly account for this bias, the supposed protective effect of moderate drinking disappears, and the risks become clear [149].

More appropriate research methods, such as Mendelian randomization (which uses genetic differences to avoid confounding), show that alcohol consumption directly increases the risk of certain diseases, rather than protecting against them [150]. These studies are less vulnerable to the biases that have affected traditional alcohol research.

> **Key Point:** When former drinkers (who often quit due to health problems) are incorrectly grouped with lifetime abstainers, it makes the abstainer group look less healthy and moderate

drinkers look healthier by comparison. Correcting for this bias eliminates the supposed benefits of moderate drinking.

3.18.2   Detailed Evidence: Mortality and Disease

All-Cause Mortality: A comprehensive review and meta-analysis of over a hundred studies, including nearly five million participants, found that after correcting for study biases, there was no reduction in the risk of death from any cause for people who drank up to about two drinks per day compared to those who never drank. In fact, the risk of dying from any cause started to increase at higher levels of drinking, and this increased risk appeared at lower levels for women than for men. For women, drinking just over two drinks per day was linked to a higher risk of death, while for men, the risk increased at about three or more drinks per day. In other words, even moderate drinking did not lower the risk of death, and higher levels clearly increased it [149].

Cardiovascular Disease: Genetic studies that avoid the usual confounding factors found that higher alcohol consumption directly increases the risk of stroke and peripheral artery disease. For example, each step up in alcohol intake was linked to about a quarter higher chance of having a stroke, and more than triple the chance of developing peripheral artery disease. There was also some evidence for a higher risk of coronary artery disease, irregular heartbeat, and abdominal aortic aneurysm, but these links were weaker. Importantly, these findings remained even after accounting for smoking and other lifestyle factors [150].

> **Key Point:** Genetic studies show that alcohol consumption directly increases the risk of stroke and peripheral artery disease. This contradicts earlier studies suggesting heart benefits from moderate drinking and shows that the supposed benefits were likely due to bias.

Cancer Risk: A large meta-analysis of over two hundred studies found that alcohol increases the risk of several types of cancer, even at levels commonly considered "moderate." For example, drinking about two standard drinks per day was linked to about three-quarters higher risk of cancers of the mouth and throat, about half higher risk of esophageal cancer, and about a third higher risk of breast cancer in women. The risk of liver, colon, rectal, and stomach cancers was also higher. Importantly, the analysis found no safe threshold, i.e., there was no level of alcohol consumption below which cancer risk was not increased [151].

> **Key Point:** There is no safe threshold for alcohol and cancer risk. Even moderate drinking increases the risk of several cancers, including breast, mouth, throat, and esophagus.

Brain Structure and Function: Recent brain imaging studies involving tens of thousands of people found that even low levels of alcohol consumption (such as one or two drinks per day) are linked to reductions in both gray and white matter in the brain. The effect of increasing from one to two drinks per day was roughly equivalent to two years of brain aging. These changes were seen throughout the brain and in both men and women [152].

> **Key Point:** Even one or two drinks per day are linked to measurable reductions in brain volume and white matter integrity, with effects similar to several years of aging.

3.18.3 Summary: Claim vs. Evidence

○ **Claim:** Drinking one or two glasses of wine per night has almost no negative effect and may provide health benefits, especially for the heart.

○ **Rationale:** Observational studies have shown that moderate drinkers often have better health outcomes than abstainers, leading to the belief that moderate alcohol consumption is protective.

○ **Evidence:** When studies correct for biases and use more rigorous methods, the supposed benefits disappear and the risks become clear:

  – **Mortality:** No reduction in risk of death for moderate drinkers; risk increases at higher levels, especially for women [149].

  – **Cardiovascular Disease:** Higher alcohol intake directly increases the risk of stroke and peripheral artery disease, and may increase the risk of other heart and vascular diseases [150].

  – **Cancer:** Even moderate drinking increases the risk of several cancers, with no safe threshold [151].

  – **Brain Structure:** One or two drinks per day are linked to reductions in brain volume and white matter, equivalent to years of aging [152].

○ **Proposed Mechanism:** The supposed benefits of moderate drinking are explained by misclassification bias, e.g., grouping former drinkers (who often quit due to health problems) with lifetime abstainers, making moderate drinkers look healthier by comparison.

○ **Conclusion:** The scientific evidence, when properly analyzed, shows that even moderate alcohol consumption is linked to increased risks of death, cardiovascular disease, cancer, and brain changes. There is no safe threshold, and the supposed benefits are artifacts of flawed study design.

– The claim that moderate drinking is harmless or beneficial is directly contradicted by rigorous scientific evidence. When biases are corrected and more robust methods are used, alcohol shows no protective effects and clear dose-dependent harm, even at low levels.

3.19   Wild vs. Farmed Salmon

Salmon is widely regarded as a healthful food, prized for its high-quality protein, omega-3 fatty acids, and micronutrients. However, the rise of aquaculture has led to persistent claims that farmed salmon is nutritionally inferior, more contaminated, or even harmful compared to its wild counterpart. Popular narratives often assert that only wild salmon is "healthy," while farmed salmon is "toxic," "full of chemicals," or "lacking in nutrients." These claims are amplified by wellness influencers, documentaries, and even some credentialed professionals, fueling confusion among consumers seeking to make informed dietary choices.

> **Key Point:** The nutritional value and safety of salmon depend on multiple factors, including species, diet, farming practices, and environmental exposures. Understanding the real differences between wild and farmed salmon requires careful examination of controlled studies, not anecdote or marketing claims.

3.19.1   Nutritional Composition

Multiple recent studies have systematically compared the nutritional profiles of wild and farmed Atlantic salmon (*Salmo salar*). Farmed salmon typically contains significantly more total fat (about three to four times as much as wild salmon) due to differences in diet and activity. For example, one study found that farmed salmon muscle contained $8.97 \pm 0.63\%$ fat, compared to $2.14 \pm 0.32\%$ in wild salmon [153]. This higher fat content means that farmed salmon delivers more calories per serving, but also more total omega-3 fatty acids per portion, even though the proportion of omega-3s in the fat is lower.

The composition of fatty acids differs as well. Farmed salmon, fed diets increasingly rich in plant oils (such as rapeseed oil), has a higher proportion of monounsaturated and omega-6 fatty acids, and a lower proportion of omega-3s, compared to wild salmon. However, the absolute amounts of the key marine omega-3s (specifically, eicosapentaenoic acid (EPA) and docosahexaenoic acid (DHA)) are similar in both types: about 520–523 mg per 100 g of fish muscle [153, 154]. This means that a typical serving of either wild or farmed salmon easily meets or exceeds recommended daily intakes of EPA and DHA.

Vitamin D content varies widely, depending on both the salmon's diet and its environment. Wild salmon can contain much higher levels of vitamin D than farmed salmon, but there is substantial variability even among wild-caught fish from different regions [155]. Farmed salmon's vitamin D content can be increased by supplementing feed, but is generally lower than the highest levels found in wild fish.

Protein content is slightly higher in wild salmon (about 16% vs. 15% in farmed), but both are excellent sources of high-quality protein and essential amino acids [154].

> **Key Point:** Farmed salmon is higher in total fat and calories, but both wild and farmed salmon provide similar amounts of omega-3s per serving. Both are excellent sources of protein and micronutrients.

3.19.2   Contaminants

A major concern raised about farmed salmon is contamination with persistent organic pollutants (POPs) such as dioxins and polychlorinated biphenyls (PCBs), as well as heavy metals like mercury and arsenic. However, recent comprehensive analyses show that wild salmon actually contains *higher* levels of these contaminants than farmed salmon. For example, wild salmon in Norway had three times the concentrations of dioxins, dioxin-like PCBs, mercury, and arsenic compared to farmed salmon, though all levels were well below European Union safety thresholds [154]. The reason is that wild salmon, feeding higher up the marine food chain, accumulates more environmental pollutants, while farmed salmon's diet is increasingly based on plant ingredients, which are lower in these contaminants.

Laboratory analyses confirm that both wild and farmed salmon are safe to eat, with contaminant levels far below regulatory limits. The shift to plant-based feeds in aquaculture has further reduced the contaminant burden in farmed salmon over the past two decades [154, 156].

> **Key Point:** Wild salmon generally contains higher levels of dioxins, PCBs, and heavy metals than farmed salmon, but both are well within safety limits for human consumption.

3.19.3   Omega-6/Omega-3 Ratio

A common argument against farmed salmon is its higher omega-6 to omega-3 ratio, due to the use of plant oils in feed. While it is true that farmed salmon has a higher omega-6/omega-3 ratio (about 0.7) compared to wild salmon (about 0.05), both are still far lower than the typical Western diet, where the ratio can exceed 15:1 [153, 154]. Both wild and farmed salmon contribute to lowering the overall dietary omega-6/omega-3 ratio, which is considered beneficial for cardiovascular health.

> **Key Point:** While farmed salmon has a higher omega-6/omega-3 ratio than wild salmon, both are far better than the typical Western diet. Including either type of salmon in your diet helps lower your overall omega-6/omega-3 ratio, supporting cardiovascular and metabolic health. The difference between wild and farmed salmon is minor compared to the benefit of eating more fish and fewer processed foods.

But why should we care about the omega-6/omega-3 ratio? The answer lies in the distinct biological roles these fatty acids play and their impact on long-term health. Omega-6 and omega-3 fatty acids are both essential polyunsaturated fats, meaning our bodies cannot synthesize them and must obtain them from the diet. However, they have different effects on inflammation and cellular function. Omega-6 fatty acids (such as linoleic acid) are precursors to molecules that can promote inflammation, while omega-3 fatty acids (such as EPA and DHA, abundant in salmon) are precursors to molecules that resolve inflammation and support cardiovascular, brain, and metabolic health [153, 154].

The balance between these two types of fats is important because they compete for the same enzymes in the body. When the diet is disproportionately high in omega-6 and low in omega-3 (as is typical in Western diets), the body tends to produce more pro-inflammatory compounds and fewer anti-inflammatory ones. This imbalance has been associated with increased risk of chronic diseases, including heart disease, certain cancers, and inflammatory conditions [153, 157].

Epidemiological and clinical studies suggest that a lower omega-6/omega-3 ratio (ideally below 4:1, and certainly below 10:1) is associated with better health outcomes, including reduced risk of cardiovascular disease, improved lipid profiles, and lower levels of inflammation [157, 158]. The modern Western diet, with ratios often exceeding 15:1, is considered far from optimal [153, 154]. In this context, both wild and farmed salmon are exceptional foods: even farmed salmon, with a ratio of about 0.7, dramatically improves the overall dietary balance when included in meals [153, 154].

Importantly, while the ratio in farmed salmon is higher than in wild salmon, both are still vastly superior to most other animal and plant foods commonly consumed in Western diets. For example, many processed foods and cooking oils have omega-6/omega-3 ratios of 20:1 or higher [157]. Thus, eating either wild or farmed salmon helps shift the overall dietary pattern toward a more favorable fatty acid balance.

> **Updated Key Point:** A lower omega-6/omega-3 ratio is linked to reduced inflammation and lower risk of chronic diseases. Both wild and farmed salmon, with their extremely low ratios (0.05 for wild and 0.7 for farmed) are far below the ideal target of 4:1 or less recommended for health. Including either type of salmon in the diet helps correct the imbalance typical of modern diets (often exceeding 15:1) and provides protective marine omega-3s that are difficult to obtain from other sources.

3.19.4   Food Authenticity and Fraud Detection

Given the price premium for wild salmon, food fraud (e.g., mislabeling farmed salmon as wild) is a real concern. Recent advances in analytical chemistry, such as MALDI-ToF mass spectrometry combined with chemometric analysis, now allow for rapid and highly accurate differentiation between wild and farmed salmon, as well as between different salmon species and geographic origins [159]. These methods achieve near-perfect classification accuracy and are being adopted to protect consumers and ensure food authenticity.

> **Key Point:** Modern analytical techniques can reliably distinguish wild from farmed salmon, helping to prevent food fraud and protect consumers.

3.19.5   Summary: Claim vs. Evidence

o **Claim:** Only wild salmon is healthy; farmed salmon is nutritionally inferior, more contaminated, and potentially harmful.

o **Rationale:** Farmed salmon is higher in fat, lower in omega-3s and vitamins, and more likely to contain pollutants.

o **Evidence:**

  – **Nutritional value:** Farmed salmon is higher in total fat and calories, but both wild and farmed salmon provide similar amounts of EPA and DHA per serving [153, 154].

  – **Vitamins:** Vitamin D content is highly variable in both wild and farmed salmon; both can be excellent sources, depending on diet and environment [155].

  – **Contaminants:** Wild salmon contains higher levels of dioxins, PCBs, mercury, and arsenic than farmed salmon, but both are well below safety limits [154, 156].

  – **Omega-6/omega-3 ratio:** Both types have favorable ratios compared to most Western foods; both are beneficial sources of omega-3s [153, 154].

  – **Food fraud:** Advanced analytical methods can accurately distinguish wild from farmed salmon, protecting consumers [159].

o **Proposed Mechanism:** Differences in diet and environment explain variations in fat content, fatty acid profile, and contaminant levels.

o **Conclusion:** Both wild and farmed salmon are safe, nutritious choices that provide high-quality protein and marine omega-3s. Wild salmon is not inherently "healthier" or "cleaner" than farmed salmon; both have strengths and trade-offs. The scientific evidence directly contradicts claims that farmed salmon is nutritionally inferior or unsafe.

- Consumers can confidently include either wild or farmed salmon in a balanced diet, focusing on overall dietary patterns rather than marketing myths.

Updated Key Point: Both wild and farmed salmon are excellent sources of protein and omega-3s, with contaminant levels well below safety limits. The choice between them can be based on taste, price, and sustainability preferences, not on unfounded health fears.

3.20   Coffee and Mold Toxins

A recurring claim in online health discussions is that coffee is contaminated with mold toxins, specifically ochratoxin A (OTA), which are said to survive roasting and brewing, attack the kidneys and bladder, and cause the body to "freak out" by dehydrating tissues and increasing urination. Some proponents go further, suggesting that these toxins are present in dangerous amounts in most coffee, that they are a major cause of kidney and bladder cancer, and that only special "clean" or "mold-free" coffees are safe to drink. This narrative is often used to market alternative coffee brands, leveraging fear of contamination and disease.

It is a well-established physiological fact that coffee passes through the body and increases urination more rapidly than water. Online marketers and influencers frequently exploit this normal, well-understood effect (pointing out that "you have to pee soon after drinking coffee") as supposed evidence that the body is urgently trying to expel toxins. In reality, this is a classic example of presenting a perfectly normal physiological response as if it were proof of harm, thereby giving false credence to their claims about coffee's dangers.

3.20.1   Scientific Analysis and Rebuttal

This claim does not withstand scientific scrutiny. The physiological reason for increased urination after drinking coffee is well established: caffeine acts as a mild diuretic by increasing blood flow to the kidneys and inhibiting antidiuretic hormone, leading to increased urine production; not because the body is "freaking out" over mold toxins. The assertion that OTA is present in dangerous amounts in most coffee is also not supported by the evidence. Regulatory agencies in Europe and globally have set strict safety limits for OTA in coffee, and multiple studies show that the vast majority of commercially available coffees contain OTA levels well below these thresholds [160].

Furthermore, the claim that OTA in coffee is a major cause of kidney or bladder disease is contradicted by large-scale meta-analyses and systematic reviews. In fact, recent evidence suggests that regular coffee consumption is associated with a *lower* risk of chronic kidney disease (CKD), end-stage kidney disease (ESKD), and CKD-related mortality [161]. While some case-control studies have suggested a possible association between coffee and bladder cancer, the more robust cohort studies do not support a causal link, even at higher levels of coffee consumption [162]. Additionally, recent narrative reviews and meta-analyses show that moderate coffee consumption is associated with reduced risk of hypertension, heart failure, atrial fibrillation, and all-cause and cardiovascular mortality, with no consistent evidence of increased risk for kidney or bladder disease [163].

Key Points:

o Caffeine, not mold toxins, is responsible for coffee's diuretic effect, i.e., increased urine production.

o OTA levels in commercial coffee are generally well below regulatory safety limits.

o Meta-analyses show coffee consumption is associated with a lower risk of CKD and kidney-related mortality.

o Cohort studies do not support a causal link between coffee and bladder cancer.

o Moderate coffee consumption is associated with reduced risk of cardiovascular and all-cause mortality.

3.20.2   Detailed Evidence: Coffee and Health Outcomes

OTA Prevalence and Safety Limits: A global systematic review and meta-analysis of 36 studies (3182 samples) found that the average OTA concentration in coffee worldwide is 3.21 μg/kg, with a pooled prevalence of 53% (i.e., OTA is detectable in about half of samples). However, the vast majority of samples are well below the European Union's maximum allowable limits: 5 μg/kg for roasted coffee and 10 μg/kg for soluble coffee. Only a small minority of samples exceeded these limits, and most commercial coffees in developed countries are far below them. The highest concentrations were found in countries with lower GDP and higher rainfall, reflecting poorer agricultural and storage practices. In most markets, OTA exposure from coffee is considered low and not a significant health risk [160].

Coffee and Kidney Disease: A 2021 systematic review and meta-analysis of 12 studies (over 500,000 participants) found that coffee consumption is associated with a significantly lower risk of developing CKD, ESKD, and albuminuria (a marker of kidney damage). The risk reduction was dose-dependent, with greater benefits seen in those drinking two or more cups per day. Specifically, coffee drinkers had a 14% lower risk of incident CKD, an 18% lower risk of ESKD, and a 28% lower risk of CKD-related mortality. The authors concluded that coffee intake is dose-dependently associated with lower risk of CKD, ESKD, and albuminuria [161].

A 2012 dose–response meta-analysis reviewed 23 case–control and 5 cohort studies on coffee and bladder cancer risk. While case–control studies suggested a small increased risk (7–29% for 1–4 cups/day), cohort studies (which are less prone to bias) found no significant association, even at four cups per day. The authors concluded that the evidence for a causal link between coffee and bladder cancer is inconclusive.

A key issue in interpreting these results is the role of confounding variables, particularly smoking. Many coffee drinkers are also smokers, and smoking is a well-established risk factor for bladder cancer. If studies do not adequately control for smoking, the increased risk observed among coffee drinkers may actually reflect the effect of smoking rather than coffee itself. This is a classic example of confounding, where the effect of one variable (smoking) is mistakenly attributed to another (coffee consumption) because the two are correlated [164]. In fact, when analyses are stratified by smoking status, the apparent association between coffee and bladder cancer disappears: no increased risk is observed among non-smokers. This strongly suggests that the observed association in some studies is due to confounding by smoking rather than a true effect of coffee.

Importantly, no dose–response relationship was observed in the more robust cohort data, and the risk was not increased in non-smokers, reinforcing the conclusion that the association seen in some studies is likely due to confounding and/or recall bias rather than a causal effect of coffee [162].

Coffee and Cardiovascular Health: A 2023 narrative review summarizes recent evidence that moderate coffee consumption is associated with decreased risk of hypertension, heart failure, atrial fibrillation, and all-cause and cardiovascular mortality. The review notes that the lowest risk is typically seen in people who drink a moderate amount of coffee (usually 1–3 cups per day), while both those who drink none and those who consume large amounts tend to have higher risks. There is no consistent evidence of increased risk for kidney or bladder disease. The authors also highlight that the method of coffee preparation (boiled vs. filtered) can affect cholesterol levels [163].

OTA and Health Risk: OTA is classified as a possible human carcinogen (Group 2B) by the International Agency for Research on Cancer, but the main health concerns are based on animal studies at much higher doses than those encountered in human diets. Regulatory agencies have set tolerable weekly intake levels for OTA, and typical coffee consumption does not approach these thresholds in most populations. Roasting and brewing reduce, but do not eliminate, OTA; however, the residual amounts in commercial coffee are generally considered safe [160].

> **Key Point:** OTA is present in about half of coffee samples worldwide, but at levels well below safety limits in most commercial products. Coffee consumption is associated with a lower risk of kidney disease and does not increase bladder cancer risk in cohort studies. Moderate coffee intake is also linked to lower cardiovascular and all-cause mortality.

3.20.3   Summary: Claim vs. Evidence

o **Claim:** Mold toxins (OTA) in coffee survive roasting and brewing, attack the kidneys and bladder, cause dehydration, and increase cancer risk. Only "mold-free" or special coffees are safe.

o **Rationale:** OTA is a mycotoxin produced by fungi in coffee beans; it is claimed to be present in dangerous amounts and to cause kidney and bladder damage.

o **Evidence:**

   – **OTA levels:** Global average OTA in coffee is 3.21 μg/kg, well below EU safety limits for most commercial products [160].

- **Kidney health:** Meta-analysis of 12 studies (over 500,000 people) found coffee drinkers have a lower risk of CKD, ESKD, albuminuria, and CKD-related mortality. The benefit is dose-dependent, with greater protection at higher intake [161].

- **Bladder cancer:** Case–control studies suggest a small increased risk, but cohort studies (the gold standard) show no association between coffee and bladder cancer, even at high intake [162].

- **Cardiovascular and overall mortality:** Moderate coffee consumption is associated with reduced risk of hypertension, heart failure, atrial fibrillation, and all-cause and cardiovascular mortality [163].

- **Mechanism:** Increased urination after coffee is due to caffeine's effect on the kidneys, not a "toxic flush" response to mold.

○ **Conclusion:** The scientific evidence directly contradicts the claim that OTA in coffee is a major health threat. OTA levels in commercial coffee are generally safe, and coffee consumption is associated with a lower risk of kidney disease, no increased risk of bladder cancer in robust studies, and reduced cardiovascular and all-cause mortality. Marketing claims about "danger coffee" or "mold-free" alternatives are not supported by the data.

- Consumers can safely enjoy regular coffee from reputable sources without fear of kidney or bladder harm from OTA. The main health effects of coffee are beneficial for kidney and cardiovascular outcomes, and the diuretic effect is due to caffeine, not mold toxins.

Updated Key Point: OTA is present in some coffee, but at levels well below safety limits in most commercial products. Coffee consumption is associated with a lower risk of kidney disease, no increased risk of bladder cancer in cohort studies, and reduced cardiovascular and all-cause mortality. Claims that coffee mold toxins are a major health threat are not supported by scientific evidence.

3.21   Coffee: Unfiltered vs. Filtered

A widely circulated claim asserts that the method of coffee preparation (specifically, whether coffee is boiled/unfiltered or filtered) has a significant impact on cholesterol levels. The argument is that boiled or unfiltered coffee (such as Scandinavian-style boiled coffee, French press, or Turkish coffee) contains much higher levels of diterpenes, which are known to raise cholesterol. In contrast, paper-filtered coffee is said to trap most of these cholesterol-raising oils, resulting in little or no effect on cholesterol. Some proponents go further, suggesting that filtered coffee may even have beneficial effects on cholesterol and cardiovascular health.

3.21.1   Scientific Analysis and Rebuttal

This claim is largely supported by the scientific literature, but with important nuances. The cholesterol-raising effect of coffee is primarily determined by the presence or absence of diterpenes (lipid-like substances found in the oil fraction of coffee beans), which are abundant in unfiltered coffee and largely removed by paper filters. Multiple large epidemiological studies, controlled trials, and mechanistic investigations confirm that regular consumption of boiled or unfiltered coffee leads to significant increases in total and LDL cholesterol, while filtered coffee has a much smaller effect, if any.

However, recent research has shown that even filtered coffee may cause a modest increase in cholesterol, particularly in women and at higher levels of consumption. The magnitude of this effect is much smaller than that seen with boiled or French press coffee, but it is not always negligible. Additionally, espresso, which is not paper-filtered but is brewed under pressure, appears to have an intermediate effect on cholesterol, with some studies showing a significant increase, especially in men.

Key Points:

o Boiled or unfiltered coffee (e.g., French press, Turkish, Scandinavian boiled) contains high levels of diterpenes, which raise cholesterol by inhibiting bile acid synthesis and reducing cholesterol breakdown in the liver.

o Paper-filtered coffee retains most of these compounds, resulting in little or no effect on cholesterol for most people, though small increases have been observed at high intakes or in certain populations.

o Espresso contains intermediate levels of diterpenes and may raise cholesterol, particularly in men and at higher consumption levels.

○ Instant coffee contains negligible diterpenes and does not raise cholesterol.

3.21.2   Detailed Evidence: Studies and Data

Mechanism and Diterpene Content: Boiled and unfiltered coffee contains about 7.2 mg of each diterpene per cup, while paper-filtered coffee retains only about 0.02 mg per cup, as the filter traps most of these cholesterol-raising oils [163]. The diterpenes inhibit bile acid synthesis, leading to reduced cholesterol breakdown and higher cholesterol [163, 165].

Epidemiological and Controlled Trials: Large population studies from Norway and the Tromsø Study have consistently shown that boiled/plunger coffee is associated with significant increases in total and LDL cholesterol for both women and men. For example, consuming 6–8 cups of boiled/plunger coffee per day increased total cholesterol by 0.20 mmol/L in women and 0.27 mmol/L in men, compared to non-drinkers [166, 167]. Filtered coffee, in contrast, was associated with a much smaller increase (about 0.10 mmol/L in women at high intakes), and no significant effect in men [166, 167].

A controlled trial by Strandhagen and Thelle found that abstaining from filtered coffee for three weeks decreased total cholesterol by 0.22–0.36 mmol/L, while consuming 600 mL (about four cups) of filtered coffee per day for four weeks raised total cholesterol by 0.15–0.25 mmol/L [165]. This demonstrates that even filtered coffee can have a modest cholesterol-raising effect, though much less than boiled coffee.

Espresso and Other Methods: Espresso, which is not paper-filtered, contains intermediate levels of diterpenes. Recent cross-sectional data from the Tromsø Study found that men drinking 3–5 cups of espresso per day had 0.16 mmol/L higher total cholesterol than non-drinkers, with a significant dose-response relationship; the effect was weaker and less consistent in women [166, 167]. Italian studies, where espresso is typically consumed in smaller volumes, have not always found this association, suggesting that cup size and preparation method matter [168, 169].

Instant coffee contains negligible amounts of diterpenes and has not been associated with increases in cholesterol in controlled studies or population data [166, 167].

Filter Quality and Variability: The effectiveness of paper filters in removing diterpenes can vary depending on filter porosity and particle size of the coffee grounds. Some studies have found that certain filters with micro-perforations allow more diterpenes to pass through, leading to small but measurable increases in cholesterol even with filtered coffee [170].

Meta-Analyses and Reviews: A meta-analysis of randomized controlled trials found that unfiltered and boiled coffee significantly increased total and LDL cholesterol compared to filtered coffee, and that the coffee dose was independently associated with the net change in cholesterol [171]. Filtered coffee exerted a smaller but still detectable increase in lipids, especially in hyperlipidemic subjects.

> **Key Point:** The cholesterol-raising effect of coffee is primarily determined by the presence of diterpenes, which are abundant in unfiltered coffee and largely removed by paper filters. Filtered coffee has a much smaller effect, but may still raise cholesterol modestly at high intakes or in sensitive individuals.

3.21.3   Summary: Claim vs. Evidence

o **Claim:** Boiled or unfiltered coffee raises cholesterol due to diterpenes; filtered coffee does not.

o **Rationale:** Paper filters trap most diterpenes, preventing their cholesterol-raising effect.

o **Evidence:**

  – **Boiled/unfiltered coffee:** Increases total and LDL cholesterol by 0.2–0.4 mmol/L at high intakes, confirmed in large population studies and controlled trials [165–167].

  – **Filtered coffee:** Small increase in cholesterol (0.07–0.15 mmol/L) at high intakes, especially in women; effect is much smaller than with boiled coffee [165–167].

  – **Espresso:** Intermediate effect; raises cholesterol in men at higher intakes, less so in women [166, 167].

  – **Instant coffee:** No significant effect on cholesterol [166, 167].

  – **Mechanism:** Diterpenes inhibit bile acid synthesis, raising cholesterol; paper filters remove most diterpenes [163, 165].

○ **Conclusion:** The scientific evidence supports the claim that boiled or unfiltered coffee raises cholesterol due to diterpenes, while filtered coffee has a much smaller effect. Espresso may raise cholesterol at higher intakes, especially in men. Instant coffee does not raise cholesterol. The impact of filtered coffee is not always zero, but is minor compared to unfiltered methods.

– For individuals concerned about cholesterol, filtered coffee is preferable to boiled or French press coffee. Those with high cholesterol or cardiovascular risk may wish to limit unfiltered coffee and moderate their intake of filtered and espresso-based coffees.

Updated Key Point: Boiled and unfiltered coffee significantly raise cholesterol due to diterpenes; filtered coffee has a much smaller effect, but may still raise cholesterol modestly at high intakes. Espresso has an intermediate effect, and instant coffee does not raise cholesterol. The brewing method is the key determinant of coffee's impact on cholesterol.

3.22   Organic vs. Conventional Foods

Many people commonly believe that organic foods are always more nutritious, completely free from pesticides and chemicals, and inherently safer or better for the environment. By contrast, there is a widespread misconception that non-organic foods are unsafe due to high pesticide residues, less nutritious, and frequently exceed legal safety limits.

3.22.1   Scientific Analysis and Rebuttal

These claims are widespread, but scientific evidence paints a more nuanced picture. While organic foods are often perceived as healthier and more environmentally friendly, the reality is more complex. Rigorous reviews and meta-analyses show that both organic and conventional foods are subject to strict safety standards, and that the differences in nutritional content and safety are generally modest. However, there are some measurable differences in composition, pesticide residues, and certain health outcomes, as well as in environmental impacts.

3.22.2   Detailed Evidence: Content and Safety

Nutritional Content: Large systematic reviews and meta-analyses have found that organic crops tend to have higher concentrations of certain antioxidants (notably polyphenols), lower cadmium levels, and a lower incidence of detectable pesticide residues compared to conventional crops. For example, a meta-analysis of 343 peer-reviewed studies found that organic crops had, on average, 19–69% higher concentrations of various antioxidants, 48% lower cadmium concentrations, and a fourfold lower frequency of pesticide residues. However, for most vitamins and minerals, the differences were small or not statistically significant. The same review found that protein and amino acid content was slightly lower in organic crops, but this is unlikely to be nutritionally significant for most consumers [172].

> **Key Point:** Organic crops are higher in some antioxidants and lower in cadmium and pesticide residues, but the overall nutritional differences are modest and not universal.

Health Outcomes: When it comes to direct health effects, the evidence is less clear. A systematic review of 35 human studies found that while organic food consumption is associated with lower pesticide exposure (as measured by urinary metabolites), there is insufficient evidence to make definitive statements about long-term health benefits. Some observational studies have reported associations between higher organic food intake and reduced incidence of certain conditions (such as infertility, birth defects, allergic sensitization, metabolic syndrome, and non-Hodgkin lymphoma), but these findings are not consistent across all studies and may be confounded by other lifestyle factors. Clinical trials comparing organic and conventional diets have generally not found significant differences in direct health outcomes or nutrient biomarkers, except for lower pesticide metabolite excretion in organic consumers [173].

> **Key Point:** Organic diets reduce pesticide exposure, but clear, direct health benefits over conventional diets have not been conclusively demonstrated in clinical trials.

Food Safety and Pesticide Residues: Both organic and conventional foods are regulated to ensure safety, and the vast majority of foods in both categories contain pesticide residues well below legal limits. Regulatory agencies in the US and Europe consistently find that only a small fraction of foods exceed maximum residue levels, and both systems are subject to rigorous standards. Organic foods do have a lower incidence of detectable pesticide residues, but "completely pesticide-free" is a myth: some residues can be present due to environmental contamination, cross-contamination, or the use of certain approved substances in organic farming [172, 173].

> **Key Point:** Both organic and conventional foods are subject to strict safety standards, and the vast majority of foods in both categories are well below legal pesticide residue limits.

Environmental Impact: A comprehensive review of 77 pairwise life cycle assessment studies found that organic cropping systems generally have lower environmental impacts than conventional systems for most indicators, including climate change, ozone depletion, ecotoxicity, human toxicity, acidification, eutrophication, resource use, water, and energy. However, organic systems tend to have lower yields (on average 22% less), and when impacts are calculated per unit of food produced (rather than per unit of land), the differences are smaller and sometimes reversed for certain crops. Notably, organic systems require more land to produce the same amount of food, and for some crops (e.g., apples, potatoes, vegetables), organic farming can have higher impacts in certain categories due to lower yields or increased use of machinery and organic fertilizers. Fertilization is a major driver of environmental impacts in both systems [174].

A global meta-analysis also found that organic farming reliably improves environmental metrics such as biodiversity, soil carbon, and profitability (due to price premiums), but yields are lower and more variable compared to conventional systems. Organic systems provide a "win-win" for environmental sustainability, while conventional systems provide more stable and higher yields [175].

> **Key Point:** Organic farming generally has lower environmental impacts and supports biodiversity, but produces lower and more variable yields, requiring more land for equivalent output.

3.22.3   Summary: Claims vs. Evidence

○ **Claim:** Organic foods are always more nutritious, completely free of pesticides, and inherently safer or better for the environment; non-organic foods are unsafe, less nutritious, and frequently exceed safety limits.

○ **Evidence:**

  – Organic crops have higher levels of some antioxidants and lower cadmium and pesticide residues, but overall nutritional differences are modest and not universal [172].

  – Both organic and conventional foods are subject to strict safety standards; the vast majority of foods in both categories are well below legal pesticide residue limits [172, 173].

  – Organic diets reduce pesticide exposure, but clear, direct health benefits over conventional diets have not been conclusively demonstrated in clinical trials [173].

  – Organic farming generally has lower environmental impacts and supports biodiversity, but yields are lower and more variable, requiring more land for equivalent output [174, 175].

○ **Conclusion:** The scientific evidence does not support absolute claims about the superiority or inferiority of either organic or conventional foods. Both are safe and nutritious, with some measurable differences in composition, pesticide residues, and environmental impact. The choice between organic and conventional foods can be informed by personal values (such as environmental sustainability or pesticide exposure), but both systems provide safe and nutritious food under current regulations.

  Key Point: Both organic and conventional foods are safe and nutritious under current regulations. Organic foods have some measurable advantages in antioxidant content and lower pesticide residues, and organic farming generally has lower environmental impacts, but the differences are modest and not absolute. Absolute claims about the superiority or danger of either system are not supported by the evidence.

3.23   Apple Cider Vinegar

Apple cider vinegar (ACV) is widely promoted as a natural remedy for a variety of health concerns. Popular claims include that ACV can aid in weight loss, lower blood sugar, improve cholesterol, enhance digestion, balance gut bacteria, relieve acid reflux, improve skin health, treat infections, and more. These assertions are common in popular media and among ACV enthusiasts, but the scientific evidence supporting many of them is limited or inconsistent.

3.23.1   Scientific Analysis and Rebuttal

The most consistent evidence for ACV's health effects relates to modest improvements in blood sugar and cholesterol, particularly in people with metabolic disorders such as type 2 diabetes or dyslipidemia (i.e., abnormal levels of lipids (fats) in the blood). Recent systematic reviews and meta-analyses provide the most comprehensive assessment to date of ACV's effects on glycemic control and lipid profiles in adults [176, 177]. Individual clinical trials have also examined ACV's effects on weight, visceral adiposity, and lipid profiles in overweight or obese subjects [178], as well as on body weight and triglycerides in obese Japanese adults [179]. (Triglycerides are a type of fat found in the blood. They come from two main sources: the fats you eat in foods like oils, butter, and meat, and from the liver, which produces triglycerides when you consume more calories than your body needs. High levels of triglycerides can increase the risk of heart disease, stroke, and other health problems. Factors that can raise triglyceride levels include obesity, an unhealthy diet, excessive alcohol consumption, and certain medications.)

> **Key Point:** The strongest evidence for ACV's benefits is for modest reductions in fasting blood sugar and total cholesterol, especially in people with type 2 diabetes or metabolic abnormalities. There is little or no robust evidence supporting claims about weight loss, digestion, skin health, or infection prevention.

3.23.2   Detailed Evidence: Reviews and Trials

Glycemic Control: A 2025 systematic review and dose-response meta-analysis, which carefully rated the quality of evidence, focused on patients with type 2 diabetes and found that apple cider vinegar (ACV) supplementation significantly reduced fasting blood sugar by about 22 mg/dL. It also lowered a marker called HbA1c by 1.53 percentage points. (HbA1c is a blood test that reflects your average blood sugar levels over the past two to three months, and is commonly used to monitor long-term glucose control in people with diabetes).

ACV supplementation also increased fasting insulin levels, but did not affect insulin resistance (a measure of how well the body responds to insulin).

The analysis found that the more ACV people consumed, the greater the reduction in fasting blood sugar, with the most pronounced effects at doses above 10 mL per day. The certainty of these findings was rated as high, and the results were consistent across different analyses [177].

A broader meta-analysis of RCTs in adults (including both diabetic and non-diabetic populations) found that ACV supplementation significantly reduced fasting plasma glucose by about 8 mg/dL and HbA1c by 0.5 percentage points compared to controls. The effect on fasting plasma glucose was more pronounced in studies lasting longer than 8 weeks and in participants with higher baseline blood sugar. No significant effects were observed on fasting insulin or insulin resistance, indicating that ACV does not consistently improve insulin sensitivity in the general population [176].

Lipid Profile: ACV consumption led to a modest but statistically significant reduction in total cholesterol (about 6 mg/dL) and a trend toward lower triglycerides (about 34 mg/dL), particularly in people with type 2 diabetes, those taking 15 mL/day or less, and those in studies lasting more than 8 weeks. No significant effects were found on LDL or HDL overall, though HDL increased slightly in non-diabetic participants. The optimal dose for lipid improvement appeared to be 15 mL/day, but meta-regression did not find a clear dose-response relationship [176].

Weight and Adiposity: A randomized clinical trial in overweight or obese adults found that ACV supplementation, combined with a calorie-restricted diet, led to greater reductions in body weight, visceral adiposity index, and improvements in lipid profile compared to diet alone [178]. Similarly, a trial in obese Japanese adults found that vinegar intake reduced body weight, body fat mass, and serum triglyceride levels [179].

Safety and Side Effects: ACV was generally well tolerated, with only minor side effects (such as stomach burning or intolerance) reported in a few studies. No serious adverse events were noted, supporting the safety of ACV as a dietary supplement in the doses studied [176, 177].

Key Points:

o ACV supplementation modestly lowers fasting blood sugar and total cholesterol, especially in people with type 2 diabetes or metabolic abnormalities [176, 177].

o The 2025 meta-analysis found a dose-response: each 1 mL/day increase in ACV reduced fasting blood sugar by 1.26 mg/dL, with greater effects above 10 mL/day [177].

o Some evidence supports modest reductions in body weight and visceral adiposity when ACV is combined with calorie restriction [178, 179].

o No significant effects on LDL cholesterol, HDL cholesterol (except a small increase in non-diabetics), fasting insulin, or insulin resistance in the general population [176].

o ACV is generally safe, with only minor side effects reported [176, 177].

o Evidence for other claimed benefits (digestion, skin health, infection prevention) is lacking or anecdotal.

3.23.3   Summary: Claim vs. Evidence

o **Claim:** ACV can aid weight loss, burn fat, lower blood sugar, improve cholesterol, help digestion, balance gut bacteria, relieve acid reflux, improve skin, treat infections, and more.

o **Rationale:** ACV is promoted as a "natural remedy" with broad health benefits, often based on anecdote or tradition rather than rigorous evidence.

o **Evidence:**

  – **Blood sugar:** ACV modestly lowers fasting plasma glucose and HbA1c, especially in people with type 2 diabetes or high baseline blood sugar [176, 177].

  – **Cholesterol:** ACV modestly reduces total cholesterol and may lower triglycerides, particularly in diabetics and at doses of 15 mL/day or less [176].

  – **Weight:** Some evidence for modest weight and fat loss when ACV is combined with calorie restriction [178, 179].

  – **Other outcomes:** No significant effects on LDL, HDL (except a small increase in non-diabetics), fasting insulin, or insulin resistance in the general population [176].

  – **Other claims:** No robust evidence supports claims about digestion, skin health, or infection prevention.

  – **Safety:** ACV is generally safe, with only minor side effects reported [176, 177].

- o **Proposed Mechanism:** ACV may slow gastric emptying, enhance glucose uptake, suppress hepatic glucose production, and stimulate bile acid excretion, but these mechanisms remain speculative and require further study [176].

- o **Conclusion:** The scientific evidence supports modest benefits of ACV for lowering fasting blood sugar and total cholesterol, especially in people with metabolic abnormalities. Other popular claims are not supported by controlled studies. ACV is safe as a dietary supplement, but should not be relied upon as a primary therapy for metabolic disorders.

  - ACV can be considered as a functional food and adjunct therapy for managing blood sugar and cholesterol, but its effects are modest and should not replace established medical treatments.
  - Further research is needed to clarify the full range of ACV's effects and to determine optimal dosing and duration.

  Updated Key Point: ACV may modestly lower fasting blood sugar and total cholesterol, especially in people with type 2 diabetes or metabolic risk, but evidence for other health claims is lacking. ACV is safe as a supplement, but its benefits are limited and should not replace proven therapies.

3.24   Food Sensitivity Tests

In recent years, food sensitivity tests have become a booming industry, promoted as a way to identify "problem foods" that supposedly cause a wide range of symptoms, from bloating and fatigue to headaches and skin issues. These tests are widely available online, in pharmacies, and even in some medical clinics, often marketed with promises of personalized nutrition and improved well-being. The most common tests measure IgG or IgG4 antibodies to dozens or even hundreds of foods, claiming that elevated levels indicate a "sensitivity" or "intolerance." Other tests use hair analysis, electrodermal screening, or cytotoxic assays, each purporting to reveal hidden food triggers [180–182].

Immunoglobulin G (IgG) is the most common type of antibody found in your blood. It plays a vital role in our immune system by helping to fight off infections. IgG4 is a specific subclass of IgG, making up a small percentage of total IgG antibodies. Both are produced in response to exposure to various substances, including foods. Most people naturally develop IgG antibodies to foods they eat regularly, and this is considered a normal immune response, not a sign of intolerance or allergy. In fact, the presence of IgG or IgG4 antibodies is more likely to indicate that our body has been exposed to a food and is tolerating it, rather than reacting negatively to it [180–184].

The rationale behind these tests is seductive: if you can identify and eliminate foods that your body "reacts" to, you can resolve chronic symptoms and optimize your health. This narrative is reinforced by testimonials and anecdotal reports, as well as by the appeal of a simple, test-based solution to complex health problems. However, the scientific evidence tells a very different story.

> **Key Point:** Most commercial food sensitivity tests, especially those based on IgG or IgG4 antibodies, are not scientifically validated and are not recommended by medical authorities. They often lead to unnecessary dietary restrictions and may delay the diagnosis of real medical conditions [180–184].

3.24.1   Scientific Analysis and Rebuttal

The claim that IgG or IgG4 antibody tests can reliably diagnose food sensitivities or intolerances does not withstand scientific scrutiny. Multiple allergy and immunology societies (including the American Academy of Allergy, Asthma & Immunology (AAAAI), the European Academy of Allergy and Clinical Immunology (EAACI), and the Canadian Society of Allergy and Clinical Immunology (CSACI)) have issued position statements explicitly warning against the use of these tests for diagnosing food intolerance or sensitivity [181–184]. The presence of IgG or IgG4 antibodies to foods is a normal physiological response that reflects exposure and tolerance, not intolerance or allergy [180–184]. In fact, higher IgG levels are often found in people who regularly eat a particular food and do not experience any symptoms [182, 184].

Controlled studies have shown that IgG-based tests have poor reproducibility, high rates of false positives, and no correlation with clinical symptoms [180, 184]. Similarly, alternative methods such as hair analysis, electrodermal testing, and cytotoxic assays lack scientific validation and are not recognized by regulatory agencies or professional societies [180, 181].

The U.S. Food and Drug Administration (FDA) does not approve any commercial food sensitivity test for diagnostic use [180].

The only validated tests for food-related adverse reactions are those used to diagnose true food allergies (IgE-mediated), such as skin prick testing, serum-specific IgE measurement, and oral food challenges conducted under medical supervision [181, 183]. For certain food intolerances (such as lactose intolerance), breath tests may be useful [180]. In contrast, "food sensitivity" is a vague and poorly defined concept, and there are no validated laboratory tests for it [180].

> **Key Point:** IgG and IgG4 antibody tests do not diagnose food sensitivities or intolerances. Their use is not supported by scientific evidence and is discouraged by major medical organizations [180–185].

3.24.2   Detailed Evidence: Statements and Studies

A position paper from the AAAAI states: "Testing for IgG or IgG4 to foods is unproven and should not be used for the diagnosis of food allergy or intolerance. The presence of specific IgG or IgG4 antibodies to foods is a normal response to exposure and does not indicate a pathologic reaction" [183]. Similar statements have been issued by the EAACI and CSACI, all emphasizing that these tests lack diagnostic value and may lead to unnecessary dietary restrictions, nutritional deficiencies, and delayed diagnosis of genuine medical conditions [181, 182, 184].

A systematic review of IgG-based food sensitivity tests found that they have poor reproducibility and do not correlate with clinical symptoms. In blinded studies, healthy individuals often test "positive" for multiple foods, despite having no symptoms when consuming them. The use of these tests can result in overly restrictive diets, increased anxiety, and, in some cases, malnutrition; especially in children [180, 182].

A 2016 review by Gocki and Bartuzi further reinforces the consensus that IgG and IgG4 antibody tests are not useful for diagnosing food allergy or intolerance. The authors explain that the presence of food-specific IgG antibodies is a normal physiological response reflecting exposure and immune tolerance, not hypersensitivity. They highlight that high levels of IgG or IgG4 are commonly found in healthy individuals and do not correlate with symptoms. The review also cites expert guidelines from major allergy societies, which explicitly advise against using IgG or IgG4 testing for food allergy diagnosis, concluding that such tests lack clinical value and should not guide dietary recommendations [185].

Alternative tests such as hair analysis, electrodermal screening, and cytotoxic assays have been evaluated in controlled studies and found to be unreliable, with results no better than chance. These methods are not recognized by any reputable medical or scientific body [180, 181].

> **Key Point:** The only validated tests for food allergies are those that assess IgE-mediated reactions (skin prick, serum IgE, oral food challenge). There are no validated laboratory tests for "food sensitivity" or "intolerance" outside of specific conditions like lactose intolerance [180, 181, 183].

3.24.3   Summary: Claim vs. Evidence

○ **Claim:** Commercial food sensitivity tests (especially IgG/IgG4-based) can identify foods that cause chronic symptoms and should be used to guide dietary choices.

○ **Rationale:** Elevated IgG or IgG4 antibodies to foods are interpreted as evidence of "sensitivity" or "intolerance."

○ **Evidence:**

– IgG/IgG4 antibodies reflect normal exposure and tolerance, not intolerance or allergy [180–184].

– Major allergy and immunology societies explicitly advise against the use of these tests for diagnosis [181–184].

– Controlled studies show poor reproducibility, high false positive rates, and no correlation with symptoms [180, 184].

– Alternative tests (hair analysis, electrodermal, cytotoxic) are not scientifically validated [180, 181].

– Only IgE-based tests and oral food challenges are validated for diagnosing true food allergies; breath tests are validated for certain intolerances [180, 181, 183].

○ **Proposed Mechanism:** IgG/IgG4 antibodies are a normal immune response to food exposure and indicate tolerance, not pathology [182, 184].

o **Conclusion:** Most commercial food sensitivity tests are not reliable or recommended. They can lead to unnecessary dietary restrictions, nutritional deficiencies, and delayed diagnosis of real medical conditions. If you suspect a food-related problem, consult a qualified healthcare professional for proper evaluation and diagnosis.

– The scientific evidence directly contradicts the claims made by the food sensitivity testing industry. Diagnosis and management of food allergies or intolerances should be based on validated methods and clinical expertise, not on unproven commercial tests [180–184].

Updated Key Point: Most commercial food sensitivity tests, especially those based on IgG or IgG4 antibodies, are not supported by scientific evidence and are not recommended by medical authorities. Only a few specific tests (for allergies or certain intolerances) are evidence-based and clinically useful. Professional medical evaluation is essential for accurate diagnosis and safe management, as IgG and IgG4 antibodies to foods reflect normal immune exposure and tolerance, not intolerance or allergy [180–185].

3.25   Vegetables Contain "Carcinogens"?

Some sources claim that vegetables are inherently dangerous because they contain dozens or even hundreds of "known carcinogens." Lists are sometimes circulated showing high numbers of identified carcinogenic compounds in common vegetables such as Brussels sprouts, mushrooms, spinach, kale, lettuce, celery, cabbage, cucumber, and broccoli. The implication is that, since these foods contain so many "carcinogens," they must be harmful, and some even go so far as to suggest that children should not eat vegetables.

3.25.1   Scientific Analysis and Rebuttal

This claim is a classic example of misunderstanding toxicology and the relationship between natural compounds, dose, and health outcomes. While it is true that plants, including vegetables, produce a wide variety of natural chemicals, some of which can be classified as "carcinogenic" in certain laboratory tests, this does not mean that eating vegetables increases cancer risk. In fact, the overwhelming body of scientific evidence shows the opposite:

higher vegetable (and fruit) intake is associated with a lower risk of cancer, cardiovascular disease, and premature death.

The key concept here is *hormesis*, which refers to the phenomenon where low doses of a potentially harmful substance can actually have beneficial effects by stimulating adaptive responses in the body. Many of the so-called "carcinogens" in vegetables are present in extremely low amounts, and their effects in the context of a whole food diet are not only harmless but may actually promote health by activating the body's natural defense systems. This is analogous to the way exercise, which causes short-term inflammation and stress, ultimately leads to improved health and resilience through repeated exposure.

A parallel can be drawn to the effects of aerobic exercise on inflammation. While a single session of exercise (i.e., *acute* exercise) can temporarily increase certain inflammatory markers in the blood, regular aerobic exercise leads to significant reductions in chronic inflammation over time. Here, "acute" means the immediate, short-term response that occurs during and shortly after one exercise session. For example, after a workout, levels of inflammatory proteins may rise briefly as part of the body's normal stress response. However, with repeated, regular exercise, the body adapts, and baseline levels of these inflammatory markers decrease, resulting in lower chronic inflammation and improved health [186].

Key Points:

- The mere presence of a "carcinogen" in a food does not mean the food is dangerous; dose and context are critical.

- Many plant compounds (that are classified as "carcinogens" only when given in extremely high doses to laboratory animals or cells) are present in vegetables at much lower levels, i.e., amounts that are not only safe for humans but may even have beneficial effects.

- The concept of *hormesis* explains how low-level exposure to natural plant chemicals or physiological stressors can strengthen the body's defenses and reduce disease risk.

- Large-scale human studies consistently show that higher vegetable intake is associated with lower (not higher) cancer risk.

3.25.2 Detailed Evidence: Reviews and Analyses

A comprehensive systematic review and meta-analysis of 95 prospective cohort studies (142 publications) involving millions of participants found that higher intake of fruits and vegetables is associated with a significantly reduced risk of cardiovascular disease, total cancer, and all-cause mortality. Specifically, for every 200 grams per day increase in fruit and vegetable intake, the relative risk of coronary heart disease, stroke, cardiovascular disease, total cancer, and all-cause mortality decreased by 8%, 16%, 8%, 3%, and 10%, respectively. The greatest reductions in risk were observed up to an intake of 800 grams per day for most outcomes, with diminishing returns at higher intakes. Notably, the study found that an estimated 7.8 million premature deaths worldwide in 2013 could be attributed to fruit and vegetable intake below 800 grams per day, if the observed associations are causal [187].

The review also examined specific types of vegetables and found that higher intake of apples/pears, citrus fruits, green leafy vegetables, cruciferous vegetables (such as broccoli, Brussels sprouts, and cabbage), and salads was associated with reduced risk of cardiovascular disease and all-cause mortality. Green-yellow vegetables and cruciferous vegetables were also linked to lower total cancer risk. These findings directly contradict the claim that vegetables are dangerous due to their natural chemical content.

The authors emphasize that the observed associations are robust across different populations and study designs, and that the benefits of vegetable consumption are likely due to a combination of fiber, vitamins, minerals, antioxidants, and phytochemicals that work synergistically to reduce oxidative stress, inflammation, and DNA damage, i.e., key processes in cancer development [187].

3.25.3 Mechanisms: Hormesis and Defense Compounds

Plants produce a wide array of natural chemicals (phytochemicals) as part of their defense against pests, pathogens, and environmental stress. Some of these compounds, such as glucosinolates in cruciferous vegetables or polyphenols in leafy greens, can be classified as "carcinogenic" in certain laboratory assays at very high doses. However, in the amounts present in a normal diet, these compounds often act as mild stressors that activate the body's own antioxidant and detoxification systems, i.e., a process known as *hormesis*.

For example, sulforaphane, a compound found in broccoli and other cruciferous vegetables, is known to induce phase II detoxification enzymes and enhance cellular defenses against oxidative damage. Rather than causing harm, these mild stressors "train" the body's cells to become more resilient, reducing the risk of chronic diseases, including cancer [187].

3.25.4   Summary: Claim vs. Evidence

○ **Claim:** Vegetables are dangerous because they contain dozens or hundreds of "carcinogens," so they should be avoided, especially by children.

○ **Rationale:** The presence of natural chemicals classified as carcinogens in laboratory tests is assumed to translate into real-world cancer risk.

○ **Evidence:**

  – The dose and context of exposure are critical; the amounts present in vegetables are extremely low and often beneficial.

  – The concept of *hormesis* explains how low-level exposure to plant defense chemicals or exercise can improve health and reduce disease risk.

  – Large-scale prospective studies and meta-analyses show that higher vegetable intake is associated with lower risk of cancer, cardiovascular disease, and premature death [187].

  – Specific vegetables (cruciferous, green leafy, green-yellow) are linked to reduced cancer and mortality risk, not increased risk.

○ **Proposed Mechanism:** Plant compounds activate the body's natural defense systems, reduce oxidative stress, and promote resilience through hormetic effects.

○ **Conclusion:** The scientific evidence directly contradicts the claim that vegetables are dangerous due to their natural chemical content. On the contrary, higher vegetable intake is consistently associated with lower cancer risk and improved health outcomes. Avoiding vegetables out of fear of "carcinogens" is not only unsupported by evidence, but may increase the risk of chronic disease.

  Key Point: The presence of natural "carcinogens" in vegetables does not make them dangerous. In fact, higher vegetable intake

is strongly associated with lower cancer risk and longer life. The
concept of *hormesis* explains how low-level exposure to plant
chemicals or exercise can strengthen the body's defenses and
promote health. The overwhelming weight of scientific evidence
supports the inclusion of a wide variety of vegetables in the diet
for cancer prevention and overall well-being.

3.26   Dietary vs. Blood Cholesterol

A common belief is that consuming foods high in cholesterol (such as
eggs, shellfish, or organ meats) will directly raise cholesterol in your blood,
leading to plaque accumulation in arteries (atherosclerosis) and increasing
your risk of heart disease and stroke. This claim is widespread in public
health messaging, online search results, and nutrition advice, where dietary
cholesterol is often conflated with blood (serum) cholesterol and assumed to
be a direct cause of arterial blockage.

3.26.1   Scientific Analysis and Rebuttal

The relationship between cholesterol in food (dietary cholesterol) and cholesterol in the blood is more complex than this claim suggests. While it is true
that high blood cholesterol (especially LDL cholesterol) is a major risk factor
for atherosclerosis and heart disease, the effect of eating cholesterol-rich
foods on blood cholesterol is generally much smaller than once thought, and
highly variable among individuals.

The body maintains cholesterol balance through a process called *homeostasis*. When dietary cholesterol intake increases, the body usually com-
pensates by producing less cholesterol in the liver. As a result, for most
people, eating more cholesterol does not lead to a large or dangerous rise
in blood cholesterol. However, this compensation is not perfect, and some
people (called *hyper-responders*) do experience a more significant increase
in blood cholesterol when eating cholesterol-rich foods.

Importantly, the effect of dietary cholesterol is amplified when the diet
is already high in saturated fat, because saturated fat impairs the function
of LDL receptors in the liver, reducing the clearance of cholesterol from the
blood. Thus, while dietary cholesterol alone is usually not the main driver
of high blood cholesterol, the combination of high saturated fat and high
cholesterol intake can be problematic in some individuals.

Recent dietary guidelines in several countries no longer recommend strict limits on cholesterol intake for most people, reflecting the evidence that moderate consumption of cholesterol-rich foods is not a major risk factor for heart disease in the general population. Instead, the focus has shifted to limiting saturated fats, which have a much stronger and more consistent effect on blood LDL cholesterol and cardiovascular risk.

Key Points:

○ Eating cholesterol-rich foods has only a modest effect on blood cholesterol for most people, due to the body's homeostatic regulation.

○ The effect can be larger in some "hyper-responders" and when saturated fat intake is high.

○ High blood (serum) cholesterol, especially LDL, is a major risk factor for atherosclerosis and heart disease.

○ The main dietary driver of high blood cholesterol is saturated fat, not dietary cholesterol.

○ Current evidence and guidelines focus on reducing saturated fat rather than dietary cholesterol for cardiovascular health.

3.26.2 Detailed Evidence: Analyses and Trials

A 2020 systematic review of 15 randomized controlled trials (over 56,000 participants, interventions lasting at least 2 years) found that reducing saturated fat intake led to a 17% reduction in the risk of combined cardiovascular events, with the greatest benefit when saturated fat was replaced by polyunsaturated fat or starchy foods. The review found little or no effect on all-cause or cardiovascular mortality, and found that the benefit was proportional to the reduction in serum cholesterol achieved. The review noted no evidence of harm (such as increased diabetes or cancer) from reducing saturated fat, and confirmed that dietary cholesterol is not the main factor driving blood cholesterol or cardiovascular risk in most people [188].

A comprehensive 2022 review on atherosclerosis and cholesterol reinforces this, noting that "the relationship between LDL cholesterol and risk for atherosclerotic cardiovascular disease (i.e., the buildup of fatty deposits, called plaque, inside the walls of arteries) is one of the most highly established and investigated issues in the entirety of modern medicine," and that "elevated LDL cholesterol is a necessary condition for atherogenesis induction." The review emphasizes that saturated fat is the primary dietary driver of LDL cholesterol, and that dietary cholesterol has only a modest effect due to homeostatic regulation. The strongest evidence for reducing cardiovascular risk is for lowering saturated fat and overall atherogenic lipoproteins, not dietary cholesterol per se [189].

> **Key Point:** The main dietary factor influencing blood cholesterol and cardiovascular risk is saturated fat, not dietary cholesterol. Dietary cholesterol has a modest effect for most people, but can be relevant in combination with high saturated fat or in "hyper-responders." High blood LDL cholesterol is a necessary driver of atherosclerosis.

3.26.3 Summary: Claim vs. Evidence

o **Claim:** Eating too much cholesterol will block your arteries, as dietary cholesterol directly builds up in your blood and causes plaque.

o **Rationale:** Public health messaging and online information often conflate dietary and blood cholesterol, implying a direct causal relationship.

o **Evidence:**

- Blood cholesterol (especially LDL) causes atherosclerosis, but dietary cholesterol contributes only modestly for most people due to homeostatic regulation [188, 189].

- The main dietary driver of high blood cholesterol is saturated fat, not dietary cholesterol.

- Reducing saturated fat intake for at least two years reduces the risk of cardiovascular events by 17%, especially when replaced with polyunsaturated fat or starchy foods [188].

- There is no evidence of harm from reducing saturated fat; no increase in cancer, diabetes, or other adverse outcomes was observed in long-term RCTs [188].

- Dietary guidelines no longer emphasize strict cholesterol limits for most people.

○ **Conclusion:** The scientific evidence does not support the claim that eating cholesterol-rich foods directly blocks your arteries for most people. The main dietary cause of high blood cholesterol and atherosclerosis is saturated fat. Dietary cholesterol is less important except for some individuals and when combined with high saturated fat intake.

Updated Key Point: Eating cholesterol-rich foods does not directly "block your arteries." The main dietary driver of blood cholesterol and artery plaque is saturated fat, not dietary cholesterol. For most people, moderate dietary cholesterol intake is not a major risk factor for heart disease; instead, focus should be on reducing saturated fat intake, as supported by large clinical trials and reviews [188, 189].

Chapter 4

What (Not) To Do

Giving advice about what (not) to do in nutrition is inherently complex. The science of nutrition is intricate, and there is no universal recipe that works for everyone. For instance, while some recommendations, such as eating more satiating foods, may benefit many people, not everyone can tolerate the same foods: some may have sensitivities or intolerances to common ingredients like potatoes or gluten, making it impossible to follow generic advice. Beyond biological differences, individual needs are shaped by a multitude of factors, including genetics, medical conditions, cultural background, past dieting experiences, and one's psychological relationship with food. Mental health, stress, access to resources, and social support also play critical roles in dietary choices and adherence. Despite these complexities, the following sections aim to distill the best available scientific evidence highlighting common themes and practical strategies that can help or hinder successful dieting, while acknowledging that all guidance must ultimately be tailored to each person's unique situation.

4.1 The Food Matrix

A core insight of modern nutrition science is that the health effects of foods cannot be understood by looking at isolated nutrients alone. Instead, nutrients exist within a "food matrix," i.e., the physical and chemical environment in which they are embedded, which fundamentally shapes digestion, absorption, metabolism, and health outcomes [190–192]. This section reviews the food matrix concept, with special attention to evidence from dairy and calcium absorption, and explains why there is no one-size-fits-all answer to nutrition.

4.1.1   What Is the Food Matrix?

The "food matrix" refers to the complex structure and organization of nutrients, non-nutrients, and bioactive compounds within foods. It encompasses not only the chemical composition of a food, but also how components are physically organized; whether nutrients are inside plant cell walls, embedded in protein-lipid networks, or present in gels or emulsions [191, 192]. The matrix determines how easily nutrients are released during digestion, how they interact with each other, and their ultimate bioavailability and physiological effects.

> **Key Point:** The health impact of a nutrient depends not just on its amount or type, but on the food matrix in which it is delivered [190, 191].

4.1.2   Why the Matrix Matters

Extensive evidence now shows that the food matrix modulates nutrient absorption and health effects in ways that cannot be predicted from nutrient content alone. This concept is especially well-studied in dairy foods and calcium, but applies across food groups [190, 193].

Dairy products are the primary source of dietary calcium in many populations, not only because of their high calcium content, but because the dairy matrix makes this calcium highly absorbable. For example, up to 40% of calcium in milk, cheese, and yogurt is absorbed, whereas the absorption from many plant foods is much lower (sometimes less than 10%) even if the total calcium content is high [193]. The difference is due to the matrix: in dairy, calcium is present in forms (such as casein micelles and calcium phosphate clusters) that solubilize well during digestion, while in certain plants, calcium is often bound to oxalate or phytate, making it insoluble and poorly absorbed [193].

The physical structure of dairy also slows gastric transit and phases the release of calcium, further improving absorption. Fermentation (as in yogurt and cheese) can increase calcium bioavailability by modifying the matrix and lowering pH [193]. In contrast, the rapid passage of calcium supplements or certain processed foods may reduce fractional absorption.

The importance of the food matrix extends beyond dairy. The structure and physical form of plant-derived foods play a vital role in determining how their nutrients are digested and absorbed, thereby influencing their metabolic effects and health outcomes. In whole fruits and vegetables, for example, the presence of intact plant cell walls serves as a barrier that limits the access of digestive enzymes to the nutrients contained within. This structural integrity slows the absorption of sugars and fats, resulting in a more gradual rise in blood glucose and lipid levels following consumption, and contributing to lower glycemic responses compared to more processed forms of the same foods [191, 192]. When fruits are juiced or pureed, the disruption of these cell walls increases the rate at which sugars become available for absorption, thereby diminishing the beneficial glycemic effects associated with the intact food matrix.

Similarly, in cereal grains, the structural organization of starch and protein within the food matrix can markedly affect the rate of carbohydrate digestion. Pasta, for instance, has a compact protein-starch matrix that tightly embeds starch granules, making them less accessible to digestive enzymes. As a result, pasta is digested more slowly and produces a smaller, more gradual increase in postprandial blood glucose compared to bread or breakfast cereals, even when the carbohydrate content is similar [192]. This difference is not simply a matter of ingredient composition, but is fundamentally rooted in the supramolecular arrangement of macronutrients within the food matrix.

Nuts and seeds provide another illustrative example. The lipids in these foods are stored within oil bodies, which are themselves enclosed by robust plant cell walls. Unless these cell walls are physically disrupted (by thorough chewing, processing, or roasting) a significant portion of the lipids remains inaccessible to digestive enzymes and thus escapes absorption. As a result, whole or minimally processed nuts tend to provide less metabolizable energy than would be predicted by their gross nutrient content, and the actual calorie value is often overestimated by conventional calculations [191, 192]. This encapsulation effect not only influences energy intake but may also have favorable effects on lipid metabolism and satiety.

Key Point: The physical form and structure of foods (whether nutrients are encapsulated within intact cell walls, embedded in a protein-starch matrix, or stored within lipid bodies) has a profound impact on the rate and extent of nutrient digestion and absorption. These matrix effects can slow carbohydrate and fat absorption, lower glycemic and lipidemic responses, and reduce the metabolizable energy of foods, underscoring the importance of considering the food matrix rather than just isolated nutrient content [191–193].

4.1.3 How the Matrix Modifies Nutrition

Several mechanisms underlie food matrix effects. First, nutrients must be released from the matrix and solubilized to be absorbed, which is a property known as bioaccessibility. In dairy, for example, calcium is solubilized by gastric acid and released from casein micelles; in spinach, much calcium remains bound as insoluble oxalate and is excreted [193]. The structure of food can also slow or accelerate digestion. Solid or gelled matrices, as found in cheese, yogurt, or whole grains, generally slow nutrient release, promoting satiety and more stable metabolic responses [192].

Nutrients in the matrix can interact, forming complexes such as calcium-phosphate or calcium-fatty acid soaps, which affect absorption and metabolism. Fermentation can generate bioactive peptides or alter mineral bioavailability [190, 193]. Foods with intact or viscous matrices often increase satiety and reduce calorie intake compared to more processed or liquid forms [191]. The matrix can also influence the rate of gastric emptying, the formation of emulsions, and the accessibility of digestive enzymes, all of which contribute to the overall metabolic impact of a food [192].

4.1.4 Implications for Dietary Guidance

Traditional nutrition advice has often focused on individual nutrients (such as saturated fat, calcium, or sugar) without considering the matrix in which they are delivered. However, the food matrix paradigm suggests that dietary recommendations should emphasize whole foods and food patterns, not just isolated nutrients [190–193].

For example, not all sources of calcium are equal: 300 mg of calcium from milk is much more bioavailable than 300 mg from spinach, due to the matrix effects of oxalate in spinach [193]. Not all sources of saturated fat are equal: saturated fat in the cheese matrix does not have the same effect on cholesterol as the same amount in butter, likely due to interactions with calcium, protein, and the slower digestion of cheese [190, 192]. Whole foods are preferable to supplements: calcium from dairy is absorbed more efficiently, and with fewer risks, than from most supplements [193].

> **Key Point:** The food matrix makes "good" or "bad" labels for individual nutrients misleading. The health effects of a nutrient depend critically on the food matrix and overall dietary pattern [190–193].

4.1.5 What (Not) To Do: Recommendations

- **Do not judge foods solely by individual nutrient content.** Always consider the whole food matrix and how processing, texture, and structure affect health impacts.

- **Do prioritize minimally processed, whole foods.** Foods with intact or gently processed matrices (e.g., whole fruits, vegetables, nuts, whole grains, fermented dairy) maximize beneficial matrix effects.

- **Do not assume all sources of a nutrient are interchangeable.** For example, calcium from dairy is more bioavailable than from most plant sources or supplements [193].

- **Do recognize the limitations of supplements.** Whenever possible, obtain nutrients from whole foods, as the matrix enhances bioavailability and synergy [193].

- **Be cautious with reductionist thinking.** Increasing or decreasing a single nutrient (like saturated fat, calcium, or sugar) may have very different effects depending on the food matrix.

> **Main Takeaway:** There is no one-size-fits-all answer in nutrition. The food matrix fundamentally shapes how our bodies respond to nutrients, meaning that "good" or "bad" categorizations of individual ingredients or molecules are often misleading. Focus on whole foods, dietary patterns, and the context in which nutrients are consumed [190–193].

4.2 Holiday Weight Gain

A widely held belief persists in both media and public health messaging that adults gain a substantial amount of weight (often cited as 5 pounds, i.e., 2.3 kg, or more) during the winter holiday period between Thanksgiving and New Year's Day. This notion has shaped public anxieties and underpins numerous dietary strategies and products marketed for "holiday weight control." However, a rigorous prospective study by Yanovski et al. provides a clear and data-driven perspective on the actual magnitude, timing, and persistence of holiday-related weight gain in adults [194]. This finding is supported by a narrative review by Díaz-Zavala et al., which found that studies in adults consistently report a significant but modest weight gain during the holiday period, typically ranging from 0.4 to 0.9 kg, with the weight often not subsequently lost [195].

4.2.1 Study Design and Methodology

To directly assess the impact of the holiday period on adult body weight, Yanovski and colleagues conducted a prospective observational study of 195 adults aged 19 to 82 years, representing a range of racial, ethnic, and socioeconomic backgrounds. Participants were recruited from the National Institutes of Health campus in Bethesda, Maryland, and were stratified by sex and other demographic variables. Importantly, no selection was made with regard to weight status, dietary habits, or prior dieting history, enhancing the generalizability of the findings to the broader U.S. adult population.

Body weight was measured on four occasions at six- to eight-week intervals: late September/early October (preholiday baseline), mid-November (pre-Thanksgiving), early or mid-January (post-New Year's), and late February/early March (postholiday follow-up). Standardized protocols ensured consistent measurement, with subjects weighed in undergarments and hospital gowns at the same time of day for each visit. Additional masking procedures (including the collection of unrelated physiological and psychological data) were employed to reduce the risk of behavior change or reporting bias related to the study's true purpose.

A subset of 165 participants returned for two further weigh-ins in June and the following September/October, enabling assessment of one-year weight trajectories.

4.2.2   Results

Contrary to popular belief, the mean weight gain during the holiday period
was modest: 0.37 kg ($\pm$1.52 kg; $P < 0.001$). This gain was statistically
significant but far below the 2–4.5 kg figures often reported in the lay press.
For comparison, weight change during the preceding "preholiday" period
was 0.18 kg ($\pm$1.49 kg; $P = 0.09$), and there was a slight, non-significant
weight loss during the postholiday period (-0.07 kg; $P = 0.36$).

Cumulatively, from late September/early October to late February/March,
the average net weight gain was 0.48 kg ($\pm$2.22 kg; $P = 0.003$). Follow-up
measurements in the subset of returning participants revealed that this
weight gain was largely maintained throughout the subsequent spring and
summer, with no significant reversal during the rest of the year. Over a
full 12-month period, the average net weight gain was 0.62 kg ($\pm$3.03 kg;
$P = 0.01$). The review by Díaz-Zavala et al. also confirms that weight
gained during the holiday period is typically not lost in the following months,
and may account for a substantial proportion of annual weight gain [195].

A notable finding was that the majority of participants experienced little
change in weight: at more than 50% of all post-baseline measurements,
weight differed from the previous measurement by less than 1 kg. Fewer than
10% of subjects gained 2.3 kg (5 lb) or more over the holiday period. There
was a trend toward greater holiday weight gain among those classified as
overweight or obese at baseline, but this did not reach statistical significance.
Attempts at weight loss during the holiday period were not associated with
significant differences in outcomes, suggesting that self-directed efforts may
be largely ineffective without structured interventions.

Self-reported perceptions of weight gain were substantially inflated compared to objective measurements: on average, participants believed they
had gained four times as much weight over the holidays as they actually had
(1.57 kg perceived vs. 0.37 kg measured; $P < 0.001$). Other studies reviewed
by Díaz-Zavala et al. report holiday weight gains in adults ranging from 0.4
to 0.9 kg, with similar findings in the US, UK, Germany, Sweden, Spain,
and Japan [195]. Some interventions focused on self-monitoring during
the holidays appear to help prevent weight gain, though the evidence base
remains limited [195].

4.2.3   Interpretation and Implications

These findings undermine the common narrative of dramatic holiday weight gain. While the average increase is small, the absence of subsequent weight loss during the spring and summer months means that even modest annual increments can accumulate over time, contributing to the progressive rise in adult body weight observed in longitudinal studies [196–198]. Previous epidemiological work demonstrates that adult weight gain of even 0.2–0.8 kg per year is typical [196, 197, 199]; the present results suggest that the holiday period may account for the majority of this annual change.

The data also indicate that individuals who are already overweight or obese may be at greater risk for larger holiday-related gains, and that subjective factors such as reductions in physical activity or increased hunger correlate with greater weight increases during this time. However, demographic variables such as age, sex, race, and socioeconomic status were not significant predictors.

Importantly, the study highlights the disconnect between perceived and actual weight change, which can drive unnecessary anxiety and maladaptive dieting behaviors. It also underscores the value of objective, longitudinal data in informing public health messaging and the development of realistic, evidence-based strategies for weight maintenance.

A particularly striking aspect of the Yanovski et al. findings is the magnitude of this disconnect: participants believed they had gained four times as much weight over the holidays as was actually measured, with a mean perceived gain of 1.57 kg versus an actual gain of 0.37 kg [194]. This overestimation is not merely a psychological curiosity; it demonstrates how subjective beliefs and expectations can diverge sharply from physiological reality, potentially influencing subsequent behavior and emotional responses. Recent experimental research has shown that such perceptions are not merely psychological, but can actively shape the body's physiological processes. For example, Crum et al. demonstrated that individuals' beliefs about the caloric content of a milkshake (even when the actual nutritional content was identical) produced distinct hormonal responses related to hunger and satiety, as measured by changes in ghrelin levels [49]. Together, these studies suggest that how we interpret and believe in our experiences (whether estimating weight gain after the holidays or judging the richness of a meal) can meaningfully influence both our psychological state and our body's metabolic responses. This underscores the importance of addressing not only behaviors, but also the mindsets and informational contexts that shape our health outcomes.

4.2.4   Scientific Context and Critique

The methodology employed by Yanovski et al. addressed several potential confounders common to self-report or retrospective studies, such as recall bias and seasonal self-monitoring. The use of direct measurements, standardized protocols, and masking of study purpose strengthens confidence in the accuracy and relevance of the findings. However, the study population, while broadly representative, was composed primarily of NIH employees, who may differ in health consciousness from the general population. As with all observational studies, the results may not generalize to all demographic groups, particularly those with extreme socioeconomic disadvantage or very different cultural practices. As Díaz-Zavala et al. note, most studies in this area use small, non-representative samples and lack long-term follow-up, which may limit the generalizability of findings [195].

Comparable studies in Europe have found similarly modest seasonal fluctuations in weight [195, 200]. The consistency of these results across settings supports the conclusion that large, rapid, and reversible holiday-related weight gains are uncommon, and that cumulative, small changes are a more significant contributor to adult obesity. Notably, Díaz-Zavala et al. found little evidence for significant holiday weight gain in children or college students, though data in these groups are limited [195].

4.2.5   What (Not) To Do: Recommendations

o **Do not panic about dramatic holiday weight gain.** The data show that, for most adults, the average gain is less than 0.5 kg and is not reversed in subsequent months. Maintaining perspective can help prevent unnecessary dietary restriction or unhealthy compensatory behaviors.

o **Do not rely solely on self-perception.** Objective measurements are essential; self-reported weight gain is often grossly overestimated, leading to undue concern and potentially counterproductive actions.

o **Do focus on year-round habits.** Since holiday period gain tends to persist, strategies for weight maintenance should emphasize consistent, sustainable behaviors rather than seasonal "detoxes" or crash diets.

○ **Be mindful of risk factors.** Individuals with overweight or obesity may be more susceptible to larger holiday gains. Monitoring activity levels and hunger cues during this period may be beneficial.

○ **Do not assume that attending more holiday events or reporting attempts to lose weight will necessarily influence outcomes.** In this study, these factors were not associated with significant differences in weight change.

Hence, the fear of substantial holiday weight gain is not supported by robust evidence. For most adults, the increase is small but persistent, and likely contributes incrementally to the gradual rise in weight observed across the lifespan. Effective prevention of adult weight gain thus requires realistic messaging, emphasis on long-term behavior change, and attention to the psychological as well as physiological aspects of eating and weight management.

> **Key Point:** Holiday-related weight gain is modest but persistent. Avoid overreacting to exaggerated claims; instead, focus on sustainable habits and objective self-monitoring throughout the year.

4.3 Night Eating and Weight Gain

The notion that eating at night inherently leads to weight gain is widespread in popular media and dieting advice. This belief has led to recommendations that individuals avoid late-night snacks or meals to prevent obesity and promote weight loss. However, the scientific evidence for a direct causal relationship between night eating and weight gain is more nuanced. A prospective study by Andersen et al. provides important data on the association between night eating habits and long-term weight change, particularly among individuals with obesity [201]. However, recent reviews indicate that the relationship between night eating and body mass index (BMI) is mixed, with some studies finding a positive association, others finding no relationship, and several reporting results that depend on factors such as age and emotional eating [202].

4.3.1   Study Design and Methodology

Andersen et al. conducted a prospective cohort study using data from
the Danish MONICA cohort, an age- and sex-stratified random sample of
the population from the western part of Copenhagen County. The study
included three examination points: an initial assessment in 1982–83, a
re-examination in 1987–88, and a third examination in 1992–93. Out of
2,987 subjects participating in 1987–88, a total of 1,050 women and 1,061
men had complete data from all three time points. Subjects who worked
night shifts were excluded to avoid confounding effects of shift work on
eating patterns and metabolism.

Night eating was assessed in 1987–88 by asking participants whether
they got up at night to eat. The primary outcomes were weight change in
the five years preceding the night eating assessment and in the six years
following it. Analyses were stratified by sex and by baseline obesity status
to explore whether the effects of night eating differed across these groups.

4.3.2   Results

Overall, 9.0% of women and 7.4% of men reported the habit of getting up
at night to eat. Among obese women, those who reported night eating
experienced a significantly greater average weight gain over the subsequent
six years (5.2 kg) compared to obese women who did not report night eating
(0.9 kg; $P = 0.004$). No significant association was found between night
eating and weight change in the five years preceding the assessment for
women, nor was there a significant association among all women regardless
of obesity status. For men, night eating was not associated with weight
change in either the preceding or subsequent periods.

4.3.3   Interpretation and Implications

The findings from Andersen et al. suggest that night eating is not universally
associated with future weight gain. Instead, the association appears to be
specific to women who are already obese at baseline. In this subgroup, night
eating may contribute to further weight gain, potentially exacerbating existing obesity. For non-obese women and for men, night eating did not predict
significant weight changes over the study period. These results indicate
that the risks associated with night eating may be context-dependent, influenced by baseline weight status and possibly by sex-specific physiological or

behavioral factors.

4.3.4   Scientific Context and Critique

The study's strengths include its prospective design, large sample size, and long follow-up period. By excluding shift workers, the authors minimized a major confounder in studies of eating timing. However, the assessment of night eating relied on a single self-reported question, which may not capture the full spectrum of night eating behaviors or distinguish between occasional and habitual patterns. Additionally, the study did not assess the total caloric intake or the nutritional quality of night-time eating episodes, which could influence the relationship between night eating and weight gain. The findings are also limited to a Danish, middle-aged population and may not generalize to other age groups or cultural contexts. The recent review by Bruzas and Allison further notes that the literature is dominated by cross-sectional studies, and that the relationship between night eating and BMI may be moderated by factors such as age and emotional eating, underscoring the need for more longitudinal research [202].

4.3.5   Summary: Claim vs. Evidence

○ **Claim:** Eating at night causes weight gain for everyone.

○ **Evidence:** Prospective data show that night eating is associated with future weight gain only among women who are already obese, and not among men or non-obese women [201]. However, the broader literature on night eating syndrome and BMI is mixed, with some studies showing a positive association, others showing no relationship, and several reporting results that depend on moderators such as age and emotional eating [202].

○ **Conclusion:** The popular claim that night eating universally leads to weight gain is not supported by evidence. Instead, the risk appears to be specific to certain subgroups, and the overall relationship remains uncertain due to mixed findings in the literature. This highlights the importance of individualized dietary recommendations and the need for more longitudinal research.

4.3.6   What (Not) To Do: Recommendations

○ **Do not assume that night eating always leads to weight gain.**
The evidence shows that night eating is not associated with future weight
gain in the general population, but may be a risk factor for further weight
gain among individuals who are already obese, particularly women.

○ **If you are already obese, be mindful of night eating habits.**
Obese women who reported night eating gained significantly more weight
over six years than those who did not.

○ **Do not generalize findings from one group to all.** The association
between night eating and weight gain was not observed in men or in
non-obese women.

○ **Consider the broader context.** Factors such as total caloric intake,
diet quality, and physical activity may play a larger role in weight
management than the timing of eating alone.

○ **Do not rely solely on self-reported eating patterns.** Objective
monitoring and a comprehensive assessment of eating behaviors provide
a more accurate picture of risk.

○ **Recognize the mixed evidence.** Recent reviews show that the
relationship between night eating and BMI is inconsistent, and more
longitudinal research is needed [202].

> **Key Point:** Night eating is not inherently linked to weight
> gain for everyone. However, for women who are already obese,
> getting up at night to eat may contribute to further weight gain.
> For others, the timing of eating appears less important than
> overall dietary habits and energy balance.

4.4   The Most Satiating Foods

A common misconception in popular nutrition is that certain foods possess
inherent "fat-burning" properties. However, the scientific consensus is that
no food directly burns fat; rather, some foods are more effective at promoting
satiety (i.e., the feeling of fullness and satisfaction after eating) which can
make it easier to maintain a calorie deficit and support weight loss. This

section reviews the evidence for the most satiating foods and examines the
scientific basis for their effects on appetite and energy intake.

4.4.1   Fibrous Fruits and Vegetables

Fibrous fruits and vegetables are characterized by low energy density (few
calories per gram) and high fiber and water content. These properties
slow digestion, increase meal volume, and promote satiety, making it easier
to control calorie intake. Intervention studies and reviews consistently
show that increasing fruit and vegetable consumption (especially in whole,
unprocessed forms) can reduce overall energy density of the diet, enhance
satiety, and support weight management. The satiety-promoting effects
are attributed to both their fiber and water content, as well as the larger
portion sizes they allow for a given calorie amount [203]. Notably, studies
demonstrate that adding substantial amounts of vegetables (e.g., 200g or
more) to meals increases satiety, and that whole fruits are more satiating
than fruit juices, likely due to their intact fiber and structure [203].

> **Key Point:** Fibrous fruits and vegetables should be a mainstay
> in the diet for appetite control and weight management, due to
> their low energy density and high fiber content.

4.4.2   Oats and High-Fiber Whole Grains

Oats are considered one of the most filling foods that are high in carbohydrates, and they score highly on measures of how well foods keep you
full. This is mainly because oats contain a special type of fiber that absorbs
water and thickens in your stomach. This thickening effect helps slow down
how quickly your stomach empties, which means you feel full for longer
after eating oats. Studies have shown that meals with oats help people feel
more satisfied, reduce hunger, and can lead to eating less later in the day
compared to cereals with less fiber. These benefits are even greater when
you eat oats that are less processed, such as old-fashioned or steel-cut oats,
and when you eat a generous serving size [204].

> **Key Point:** Oats, due to their high soluble fiber content and
> ability to absorb water, are among the best foods for promoting
> satiety and preventing overeating.

4.4.3   Fish and Lean Proteins

Fish emerges as the most satiating option among protein-rich foods. In controlled studies where participants consumed equal-calorie portions of beef, chicken, and fish, fish consistently resulted in greater post-meal satiety; people reported less hunger and ate less at subsequent meals compared to other protein sources [205]. This pronounced effect may be due to fish's distinctive amino acid profile, its influence on serotonin signaling, and its relatively slower rate of digestion. Notably, the Satiety Index study ranked fish higher than all other protein-rich foods, emphasizing its unique capacity to regulate appetite [206].

Despite these findings, many people feel that fish is less filling in everyday settings. This apparent contradiction arises largely because fish is often consumed in smaller, lower-calorie portions than beef or chicken, leading to a lower total energy intake and, consequently, less perceived fullness. Additionally, psychological, cultural, and sensory factors (such as meal expectations, the visual impact of portion size, and personal preferences) play a significant role in how satisfying a meal feels. The method of preparation and the texture of fish may further influence its satiating power. Thus, the gap between research results and real-world experience reflects differences in serving size, preparation, and context, rather than a flaw in the underlying science.

> **Key Point:** Fish is the most satiating protein source in scientific studies, making it an excellent choice for appetite control and weight management, especially when consumed in adequate portions.

4.4.4   Boiled Potatoes

Boiled potatoes top the satiety index, outperforming all other tested foods, including protein- and fiber-rich options. Their high satiety is attributed to a combination of factors: high water content, moderate fiber, low energy density, and the presence of a unique protein called proteinase inhibitor II (PI2). PI2 has been shown in animal and human studies to increase circulating levels of a hormone that suppresses appetite and slows gastric emptying, thereby enhancing feelings of fullness [207]. The Satiety Index study found that boiled potatoes were more than three times as satiating as white bread, and significantly more satiating than other carbohydrate-rich foods [206]. Additional research demonstrates that potato-derived protease inhibitors can reduce food intake and body weight gain in animal models, supporting their potential role in appetite regulation [207].

> **Key Point:** Boiled potatoes are the most satiating food tested, likely due to their unique combination of water, fiber, and PI2 content, making them a powerful tool for appetite control.

4.4.5 Summary and Practical Implications

The foods with the strongest evidence for promoting satiety (namely, fibrous fruits and vegetables, oats, fish, and boiled potatoes) share common characteristics: low energy density, high fiber or protein content, and, in some cases, unique bioactive compounds that enhance satiety hormone release. Incorporating these foods into the diet can help reduce hunger, support adherence to a calorie deficit, and facilitate weight management.

> **Key Point:** While no food directly "burns fat," choosing foods with high satiety value can make it easier to eat less without feeling deprived, supporting sustainable weight loss.

4.4.6 What (Not) To Do: Recommendations

- **Do prioritize foods with high satiety index scores.** These include boiled potatoes, fish, oats, and fibrous fruits and vegetables.

- **Do focus on whole, minimally processed forms.** Whole fruits and vegetables, intact whole grains, and simply prepared proteins maximize satiety.

- **Do not rely on the idea of "fat-burning" foods.** The key to weight loss is maintaining a calorie deficit, which is easier when hunger is controlled.

- **Be mindful of food form and preparation.** Juices, fried foods, and highly processed grains are less satiating than their whole-food counterparts.

4.5 Saturated Fats in the Diet

Saturated fats are found mainly in animal products like butter, cheese, red meat, and whole milk, but also in a few plant-based foods such as coconut oil and palm oil [208, 209]. Saturated fats have long been a focus of dietary guidelines, with most recommendations advising people to limit their intake to less than 10% of daily calories. However, the real-world relationship between saturated fat and health is more nuanced than "good" or "bad." Recent research highlights that not all saturated fats are the same, and their health effects depend greatly on the food source, your overall diet, and individual risk factors [208–211].

4.5.1 Where Do Saturated Fats Come From?

Recent research using national survey data from the United States looked at what foods contribute the most saturated fat to people's diets. The study included nearly 12,000 participants of all ages, who reported everything they ate in a 24-hour period. Researchers grouped foods into categories like dairy, meats, seafood, plants, and an "other" group (which included processed foods, snacks, and baked goods). They then calculated what percentage of total saturated fat came from each group.

The results showed that dairy products (including milk, cheese, yogurt, pizza, and ice cream) made up about 28% of all saturated fat eaten, while meats (including red meat, poultry, and processed meats) contributed about 22%. Plant sources (like coconut and palm oil) and seafood made up much smaller shares, at 7.5% and 1.2% respectively. The largest single category was "other," which included processed foods and snacks, accounting for about 42% of saturated fat intake. When looking at specific foods, unprocessed red meats and sweet bakery products (like cakes and cookies) were the top contributors, followed by cheese, milk, and pizza. These patterns were similar across different age, sex, and income groups, though children got more saturated fat from dairy and adults got more from meat [208].

> **Key Point:** Most saturated fat in the US diet comes from dairy, meat, and processed foods, with unprocessed red meat and sweet baked goods as leading sources [208].

4.5.2   Does the Source of Saturated Fat Matter?

To understand if the health effects of saturated fat depend on where it comes from, researchers have used both large population studies and clinical trials. In population studies, people's diets are tracked over years, and their health outcomes (like heart disease) are recorded. These studies have found that saturated fat from dairy foods (milk, cheese, yogurt) is not linked to a higher risk of heart disease, and in some cases, may even be linked to a lower risk. In contrast, saturated fat from red and processed meats is more consistently linked to higher risk of heart disease and other chronic illnesses [208, 210].

Clinical trials, where people are assigned to eat certain foods for weeks or months, have also shown that the "food matrix"—the combination of nutrients and structure in a food—matters. For example, several trials compared cheese and butter, which have similar amounts of saturated fat. They found that eating cheese had a more neutral or even beneficial effect on cholesterol levels compared to butter. This may be because cheese contains calcium, protein, and is fermented, which can change how the body processes fat [211].

> **Key Point:** Saturated fat from dairy is not linked to higher heart disease risk, while saturated fat from red and processed meats is more concerning. The food matrix matters [208, 210, 211].

4.5.3   Regular-Fat vs. Low-Fat Dairy

For many years, health guidelines have recommended low-fat dairy over regular-fat (full-fat) dairy, assuming that less saturated fat is always better. However, recent reviews and expert panels have looked at all the available evidence, including large population studies and clinical trials. These reviews found that eating regular-fat dairy does not increase the risk of heart disease compared to low-fat dairy. In clinical trials, both types of dairy had similar effects on cholesterol, blood pressure, inflammation, and other markers of heart health. This was true for adults, and the limited studies in children also found no difference in risk between regular-fat and low-fat dairy [210, 211].

The studies included both short-term trials (weeks to months) where people ate more or less dairy, and long-term studies following people for years. Across these studies, neither regular-fat nor low-fat dairy was linked to worse heart health. In fact, some studies found that children who drank whole milk were less likely to be overweight than those who drank low-fat milk.

> **Key Point:** There is no evidence that regular-fat dairy is worse for heart health than low-fat dairy. Both are neutral for cardiovascular risk in adults, and likely in children [210, 211].

4.5.4   Why the Confusion?

A major reason for confusion is that "saturated fat" is defined and measured in different ways. Some guidelines and studies group all foods with saturated fat together, without considering the actual amount or type of fatty acids, cholesterol, or how the food is processed. For example, lard is often called a "saturated fat," but it actually contains more unsaturated than saturated fatty acids [209]. Some definitions are based on whether a fat is solid at room temperature, but this is not a reliable way to judge health effects, since melting point does not match up with fatty acid content or health impact.

Studies also often do not specify what foods are being compared or what replaces saturated fat in the diet. The health impact of reducing saturated fat depends on what you eat instead: swapping saturated fat for unsaturated fat (from nuts, seeds, or plant oils) is beneficial, but replacing it with refined carbohydrates or sugar does not improve health and may even make things worse [210].

> **Key Point:** Definitions of "saturated fat" are inconsistent and often misleading. The health impact depends on the food source and what replaces saturated fat in the diet [209, 210].

4.5.5   Summary and Practical Implications

The evidence from both population studies and clinical trials does not support the idea that all saturated fats are equally harmful, or that regular-fat dairy should be avoided. Instead, the source and context of saturated fat matter. Dairy saturated fat is not a health risk for most people, while saturated fat from red and processed meats and from processed foods is more concerning. The focus should be on overall dietary patterns and food quality, not just saturated fat numbers.

Key Point: The health impact of saturated fat depends on its source and the overall diet. Regular-fat dairy is not a health risk for most people. Focus on food quality and healthy swaps, not just saturated fat numbers [210, 211].

4.5.6   What (Not) To Do: Recommendations

○ **Do not fear regular-fat dairy.** Enjoy milk, cheese, and yogurt in moderation as part of a balanced diet.

○ **Do focus on the whole diet.** Prioritize minimally processed foods, plenty of vegetables, whole grains, and healthy fats from plants and fish.

○ **Do limit processed meats and fatty cuts of red meat.** These are more strongly linked to health risks than dairy.

○ **Do not rely on food labels alone.** Consider the overall quality and context of the food, not just its saturated fat content.

○ **Do replace saturated fat with unsaturated fat when possible.** Use olive oil, nuts, or avocado instead of butter or lard for cooking.

○ **Do not replace saturated fat with sugar or refined starches.** This does not improve health and may make things worse.

> **Main Takeaway:** Not all saturated fats are equal. Regular-fat dairy is not a health risk for most people. Focus on food quality, the source of saturated fat, and healthy dietary patterns rather than just saturated fat numbers.

4.6   Unsaturated Fats in the Diet

Unsaturated fats are a broad category of dietary fats that are generally considered beneficial for health, especially when they replace saturated or trans fats in the diet. Unlike saturated fats, which have no double bonds between carbon atoms, unsaturated fats contain one or more double bonds, which influence their physical properties and effects in the body. Unsaturated fats are typically liquid at room temperature and are found mainly in plant-based foods and some types of fish. They are divided into two main types: monounsaturated fats (MUFAs) and polyunsaturated fats (PUFAs), each with distinct sources and health effects [212, 213].

4.6.1 Where Do Unsaturated Fats Come From?

Unsaturated fats are abundant in a variety of plant-based foods and certain fish. Major sources of monounsaturated fats include olive oil, canola oil, avocados, and many nuts (such as almonds, peanuts, and cashews). Polyunsaturated fats are found in high amounts in vegetable oils (such as soybean, sunflower, safflower, and corn oil), walnuts, flaxseeds, chia seeds, and fatty fish (such as salmon, mackerel, sardines, and trout). Omega-3 and omega-6 fatty acids are two important types of polyunsaturated fats, with omega-3s especially concentrated in fatty fish and some seeds [212].

> **Key Point:** Unsaturated fats are mainly found in plant oils, nuts, seeds, avocados, and fatty fish, with different foods providing different types of unsaturated fats.

4.6.2 Health Effects of Unsaturated Fats

A large body of evidence from population studies and clinical trials shows that replacing saturated fats with unsaturated fats (especially polyunsaturated fats) lowers LDL ("bad") cholesterol and reduces the risk of heart disease [213, 214]. For example, a major review of randomized controlled trials found that when people increased their intake of polyunsaturated fats (PUFAs) in place of saturated fats, their risk of coronary heart disease (CHD) events dropped by 19%. This means that for every 5% increase in calories from PUFAs (instead of saturated fats), the risk of heart disease fell by about 10%. The benefits were even greater in studies that lasted longer, suggesting that the positive effects of unsaturated fats build up over time [212].

Monounsaturated fats, such as those in olive oil and avocados, are a key component of the Mediterranean diet, which is linked to lower rates of cardiovascular disease and improved metabolic health. A large meta-analysis of over 800,000 people found that higher intakes of olive oil (a major source of MUFA) were associated with an 11% lower risk of death from any cause, a 12% lower risk of dying from heart disease, a 9% lower risk of cardiovascular events, and a 17% lower risk of stroke. However, these benefits were mainly seen with olive oil, not with MUFAs from animal sources [213].

Polyunsaturated fats, particularly omega-3 fatty acids from fish, have additional benefits for heart rhythm, blood pressure, and inflammation.

Some studies suggest that higher intake of unsaturated fats may also help with weight management and reduce the risk of type 2 diabetes [212].

Importantly, recent research shows that replacing saturated fats with unsaturated fats (either MUFA or PUFA) does not increase levels of lipoprotein(a), a blood marker linked to heart disease risk. In contrast, replacing saturated fat with refined carbohydrates or trans fats can actually raise lipoprotein(a) levels, which is not desirable [214].

> **Key Point:** Replacing saturated fats with unsaturated fats, especially from plant oils and fish, lowers heart disease risk and improves cholesterol levels.

4.6.3 Monounsaturated vs. Polyunsaturated Fats

Both monounsaturated and polyunsaturated fats are beneficial, but they have slightly different effects. Monounsaturated fats (MUFAs) are especially prominent in olive oil and avocados, and are associated with improved cholesterol profiles and reduced inflammation. However, the health benefits of MUFAs seem to depend on their source. The largest benefits are seen with olive oil, which is a staple in Mediterranean diets. In fact, only olive oil (not MUFAs from animal sources) was consistently linked to lower risk of death, heart disease, and stroke in large population studies [213].

Polyunsaturated fats (PUFAs) include both omega-6 and omega-3 fatty acids. While omega-6 fats (from vegetable oils and nuts) are essential and beneficial in moderation, omega-3 fats (from fatty fish, flaxseed, and walnuts) are particularly important for heart and brain health. Most people consume enough omega-6, but may need to increase omega-3 intake for optimal health [212].

> **Key Point:** Both MUFAs and PUFAs are healthy, but omega-3 PUFAs from fish and seeds are especially important for heart and brain health.

4.6.4 Why the Emphasis on Unsaturated Fats?

The health benefits of unsaturated fats are not just due to what they provide, but also what they replace. Studies show that swapping saturated or trans fats for unsaturated fats leads to better health outcomes, while replacing saturated fat with refined carbohydrates does not provide the same benefit.

For example, a recent meta-analysis found that replacing saturated fat with carbohydrates or trans fats actually increased levels of lipoprotein(a), a risk factor for heart disease, while replacing saturated fat with unsaturated fats did not [214].

The Mediterranean and traditional Asian diets, which are rich in unsaturated fats from plant oils, nuts, and fish, are consistently linked to lower rates of chronic disease. In these diets, the main sources of unsaturated fats are minimally processed plant oils (like olive oil), nuts, and fatty fish, rather than processed foods [212, 213].

> **Key Point:** The benefit of unsaturated fats is greatest when they replace saturated or trans fats, not when they are simply added to a diet high in refined carbs or processed foods.

4.6.5   Summary and Practical Implications

The evidence strongly supports the inclusion of unsaturated fats from plant oils, nuts, seeds, and fish as part of a healthy diet. These fats improve cholesterol, reduce inflammation, and lower the risk of heart disease. The focus should be on food quality and dietary patterns, not just fat percentages. Notably, the greatest benefits are seen when unsaturated fats come from sources like olive oil and fatty fish, and when they replace less healthy fats in the diet [212–214].

> **Key Point:** Unsaturated fats are a cornerstone of healthy eating patterns. Prioritize plant oils, nuts, seeds, and fatty fish for optimal heart and metabolic health.

4.6.6   What (Not) To Do: Recommendations

○ **Do use plant oils (like olive, canola, or sunflower) for cooking and dressings.**

○ **Do eat a variety of nuts and seeds regularly.**

○ **Do include fatty fish (such as salmon, sardines, or mackerel) in your diet at least once or twice a week.**

○ **Do not rely on processed foods labeled "vegetable oil" without checking for trans fats or hydrogenated oils.**

○ **Do not assume all plant oils are equally healthy**; focus on minimally processed oils and avoid those high in trans fats.

○ **Do replace saturated and trans fats with unsaturated fats,** not with refined carbohydrates or sugars.

> **Main Takeaway:** Unsaturated fats from plant oils, nuts, seeds, and fish are beneficial for heart and metabolic health, especially when they replace saturated or trans fats in the diet.

4.7 Fats: Sorting Out the Confusion

After reviewing the sources, health effects, and practical recommendations for both saturated (Section 4.5) and unsaturated (Section 4.6) fats, it is important to address the confusion surrounding the different types of dietary fats and to clarify how they are related and what their implications are for health.

Dietary fats are among the most misunderstood nutrients in nutrition science. The term "fat" encompasses a wide variety of molecules with different chemical structures, physical properties, and health effects. Understanding the distinctions between saturated, unsaturated, trans, monounsaturated, and polyunsaturated fats is essential for making informed dietary choices. This section clarifies what each type of fat is, how they are related, and what the evidence says about their health impacts.

4.7.1 What Are Fats? The Basics

Fats, or fatty acids, are molecules made up of long chains of carbon atoms bonded to hydrogen atoms. The main categories of fats are defined by their chemical structure. The differences between types of fats come down to the presence and configuration of double bonds between carbon atoms. Saturated fats have no double bonds and are not subdivided for practical dietary purposes. Unsaturated fats, which have at least one double bond, can be further divided into monounsaturated and polyunsaturated fats, depending on the number of double bonds. Both of these subtypes are usually found in the cis configuration in natural foods. However, unsaturated fats can also exist in the trans configuration, forming trans fats. Thus, trans fats are technically a subtype of unsaturated fat, but their unique configuration makes them behave differently in the body [215–217].

○ **Saturated fats**: These have no double bonds; all carbon atoms are "saturated" with hydrogen. They are typically solid at room temperature and found in animal products (butter, cheese, red meat) and some tropical oils (coconut, palm). Saturated fats are one of the two main categories of fats, defined by the absence of double bonds in their carbon chains. In standard nutrition science, saturated fats are not further subdivided into practical subcategories, though biochemically they can be classified by chain length. For dietary guidance, they are treated as a single group [215–217].

○ **Unsaturated fats**: These have one or more double bonds in their carbon chains and are usually liquid at room temperature. Unsaturated fats are the broad category that includes several subtypes:

 – **Monounsaturated fats (MUFAs)**: A subtype of unsaturated fats with exactly one double bond, usually in the cis configuration (hydrogens on the same side). Found in olive oil, avocados, and many nuts. MUFAs are generally considered beneficial for heart health [215–217].

 – **Polyunsaturated fats (PUFAs)**: A subtype of unsaturated fats with two or more double bonds, usually in the cis configuration. Includes omega-3 and omega-6 fatty acids, found in fatty fish, walnuts, flaxseed, and many vegetable oils. PUFAs are essential fats and are also considered beneficial [215–217].

 – **Trans fats**: A special configuration of unsaturated fat in which at least one double bond is in the trans configuration (hydrogens on opposite sides), making the molecule straighter and more similar in shape to saturated fat. Trans fats can be monounsaturated or polyunsaturated, and may be naturally occurring (in small amounts in dairy and meat from ruminant animals) or industrially produced by partial hydrogenation of vegetable oils. All trans fats are unsaturated by chemical definition, but their physical properties and health effects are more similar to or worse than saturated fats [215–218].

 Key Point: All trans fats are unsaturated fats by chemical definition, but their "trans" configuration gives them physical properties and health effects more similar to saturated fats [215–217].

4.7.2 Which Fats Are Harmful? Which Are Beneficial?

Trans fats, especially those produced industrially, are the most harmful type of fat. Strong evidence from metabolic studies, clinical trials, and large population studies shows that trans fats raise LDL ("bad") cholesterol, lower HDL ("good") cholesterol, increase inflammation, and substantially increase the risk of coronary heart disease (CHD) [215–219]. A meta-analysis found that just a 2% increase in energy intake from trans fats is associated with a 23% increase in the risk of heart disease [215–219]. This risk is higher than for any other type of fat.

Saturated fats also raise LDL cholesterol, but their effects depend on the food source and the overall dietary pattern. Saturated fat from red and processed meats is more strongly linked to heart disease than saturated fat from dairy [215]. The health impact of saturated fat is less severe than that of trans fat, but replacing saturated fat with unsaturated fat (especially polyunsaturated) is beneficial [215, 217].

Unsaturated fats in the cis configuration (cis-MUFAs and cis-PUFAs) are generally beneficial. Replacing saturated or trans fats with unsaturated fats lowers LDL cholesterol and reduces the risk of heart disease [215, 217]. Polyunsaturated fats (especially omega-3s) have additional anti-inflammatory and heart-protective effects.

Natural (ruminant) trans fats are present in small amounts in dairy and meat from cows and sheep. Their intake is much lower than industrial trans fats, and current evidence suggests their impact on heart disease risk is minor at typical intake levels [215, 217]. However, gram for gram, all trans fats (industrial or ruminant) have similar adverse effects on blood lipids [217].

> **Key Point:** Industrial trans fats are the most harmful; saturated fats are less harmful but should be limited; unsaturated fats (especially from plants and fish) are beneficial. All trans fats are unsaturated, but not all unsaturated fats are healthy [215–217].

4.7.3 What (Not) To Do: Recommendations

○ **Do avoid industrial trans fats completely.** Check labels for "partially hydrogenated oils" and avoid foods containing them, even if the label says "0g trans fat" (small amounts per serving can add up) [215,216].

○ **Do limit saturated fat, especially from red and processed meats.** Choose lean meats, fish, and plant-based proteins more often [215].

○ **Do replace saturated and trans fats with unsaturated fats.** Use olive oil, canola oil, nuts, seeds, and fatty fish as primary fat sources [215, 217].

○ **Do not assume all unsaturated fats are healthy.** Trans fats are unsaturated but are the most harmful type [215–217].

○ **Do not worry about small amounts of natural trans fats from dairy and meat if your overall diet is healthy.** Their intake is low and not a major health concern at typical levels [215, 217].

○ **Do not use CLA (conjugated linoleic acid) supplements for health.** They are a type of trans fat and have similar adverse effects on blood lipids [217].

> **Main Takeaway:** Not all fats are created equal. Trans fats (especially industrial) are the most harmful and should be avoided entirely. Saturated fats are less harmful but best limited, especially from processed meats. Unsaturated fats from plants and fish are beneficial and should be the main source of dietary fat. Always check food labels and prioritize whole, minimally processed foods for optimal health [215–217].

4.8   Portion Size, Not Plate Size

A popular weight loss tip circulating in recent years is the recommendation to use smaller plates to reduce food intake. The rationale is that larger plates encourage larger servings, which in turn may lead to overeating due to visual cues and the psychological urge to "clean your plate." Proponents often claim that dinner plates today are significantly larger than in previous decades, and that simply switching to smaller dishware can help control portions and promote weight loss. But does the scientific evidence support this advice?

4.8.1   What Do Studies Show?

The idea that plate size influences how much we eat is rooted in the observation that people tend to serve themselves more food when using larger plates, a phenomenon sometimes attributed to visual illusions such as the Delboeuf illusion [220]. Early experimental studies suggested that people might serve and consume more food when given larger plates or bowls, especially in buffet or party settings. However, when it comes to actual energy intake at meals, the evidence is far less clear.

A series of well-controlled laboratory experiments by Rolls et al. directly tested whether using smaller plates reduces how much people eat at a meal. Across three crossover studies, participants were served identical foods on plates of varying sizes (ranging from 17 cm to 26 cm in diameter) and allowed to eat as much as they wanted. The results were consistent: plate size had no significant effect on the amount of food consumed. Whether participants served themselves from a large dish, received a fixed portion on different-sized plates, or selected foods from a buffet, their energy intake remained essentially unchanged regardless of plate size. When smaller plates were used, participants simply made more trips to the buffet or adjusted their behavior to compensate, resulting in no meaningful reduction in total intake. Hunger and food taste were rated as much more important influences on intake than plate size [221].

Other research has also examined whether specific plate design features, such as rim width and rim coloring, can influence perceived portion size via visual illusions. In a pair of within-subjects experiments, McClain et al. tested whether the Delboeuf illusion applies to food on a plate by manipulating plate rim width and rim coloring in photographic images [220]. Adult participants viewed side-by-side images of plates with varying rim designs and food portion sizes, and were asked to select which plate appeared to contain more food.

In Experiment 1, wider plate rims led participants to overestimate food portion diameter by about 5% and visual area by 10% compared to plates with very thin rims. This effect was more pronounced with larger food portions. In Experiment 2, plates with colored rims (e.g., solid blue or blue lines) caused participants to overestimate food portion diameter by 1.5% and area by 3% compared to plates with no rim coloring, with the effect being greater for smaller food portions. Food type did not significantly affect these perceptions. The authors concluded that plate rim width and coloring can exaggerate perceived food portion size, suggesting that plate design may influence how much food people *think* they are eating, even if it does not necessarily reduce actual intake [220].

Key Point: In controlled laboratory settings, using smaller plates does *not* reliably reduce how much people eat at a meal. However, certain plate designs (such as wider or colored rims) can make portions appear larger, potentially influencing perceived satiety or satisfaction, though not necessarily actual intake.

4.8.2   Systematic Reviews and Meta-Analyses

To assess the broader evidence, Robinson et al. conducted a systematic review and meta-analysis of studies that experimentally manipulated dishware size and measured subsequent food intake. The review included nine experiments and found that most studies reported no significant difference in food intake when participants ate from smaller versus larger dishware. The overall effect, if present, was small and inconsistent, with substantial heterogeneity across studies. The authors concluded that the evidence does not support a consistent or meaningful effect of dishware size on energy intake, and that recommendations to use smaller plates as a public health strategy may be premature [222].

Key Point: The best available evidence from systematic reviews shows that smaller plates have, at most, a marginal and inconsistent effect on food intake. Most studies find no significant difference.

4.8.3   Why the Disconnect?

Some earlier studies and popular articles cite research showing that people serve themselves more food when using larger plates or bowls, especially with amorphous foods or in social settings. For example, Wansink and colleagues found that people served themselves more ice cream when given larger bowls and spoons. However, these studies often measured the amount served, not the amount actually eaten, and were conducted in less controlled environments such as parties or buffets [223]. In contrast, when intake is measured precisely in laboratory settings, people appear to regulate their consumption based on hunger, satiety, and habitual portion sizes, rather than plate size alone [221].

It is possible that plate size may have a greater effect in real-world settings where distractions, social cues, or lack of attention to hunger signals are more prominent. However, even in these contexts, the effect is likely to be small compared to other factors such as portion size, food palatability, and energy density.

4.8.4 Portion Size vs. Plate Size

Importantly, while plate size itself does not appear to drive intake, the portion size of food served *does* have a robust and sustained effect on how much people eat. Numerous studies show that when people are served larger portions, they consume more food, often without realizing it, and this effect persists over days or weeks [224, 225]. Thus, strategies that focus on reducing portion sizes (such as serving less food, using pre-portioned packages, or increasing the proportion of low-energy-dense foods) are more effective for controlling intake than simply switching to smaller plates.

> **Key Point:** Reducing portion size is a proven strategy for lowering energy intake. Plate size alone, without a change in portion size, is unlikely to have a meaningful impact.

4.8.5 What (Not) To Do: Recommendations

○ **Do not rely on smaller plates as a primary weight loss strategy.** The evidence shows little to no effect on actual food intake in controlled settings.

○ **Do focus on portion size.** Serve smaller portions, use pre-portioned foods, or increase the proportion of low-calorie, high-fiber foods on your plate.

○ **Do pay attention to hunger and satiety cues.** Mindful eating and awareness of internal signals are more effective than visual tricks.

○ **Be cautious with "hacks" that promise easy results.** Sustainable weight loss is best achieved through evidence-based changes in eating habits and food environment.

○ **Do recognize that environmental cues matter, but prioritize those with proven effects.** Portion size, food variety, and energy density are more influential than plate size.

> **Main Takeaway:** Using smaller plates is not a reliable or effective strategy for reducing food intake or promoting weight

loss. Focus instead on controlling portion sizes, choosing lower energy-dense foods, and paying attention to hunger and fullness cues.

4.9   Scale Weight vs. Body Composition

A pervasive misconception in weight management is the belief that the number on the scale is the ultimate measure of progress, health, and physical appearance. This scale-centric mentality has led countless individuals to pursue rapid weight loss through extreme caloric restriction, often at the expense of muscle mass and metabolic health. However, the distinction between losing weight and losing fat represents one of the most important concepts in evidence-based nutrition and fitness. Understanding body recomposition (i.e., the process of simultaneously losing fat while maintaining or gaining muscle) can fundamentally change how we approach health and aesthetic goals.

4.9.1   What Is Body Recomposition?

Body recomposition refers to the process of changing body composition by reducing fat mass while maintaining or increasing lean muscle mass, often with minimal change in total body weight. This concept challenges the traditional weight-loss paradigm that equates success with decreasing numbers on a scale. Two individuals can have practically the same weight yet dramatically different body shapes and health profiles, depending on their ratio of muscle to fat tissue.

The fundamental principle underlying body recomposition is that muscle and fat tissue have vastly different characteristics: five pounds of fat occupies significantly more space on the body than five pounds of muscle, due to differences in tissue density. Muscle tissue is approximately 18% more dense than fat tissue, meaning that as individuals lose fat and gain muscle, they may see dramatic improvements in body shape, clothing fit, and physical performance while experiencing little to no change in scale weight.

It is important to note that the number on the scale does not distinguish between fat mass, muscle mass, bone, or water content. Thus, relying solely on scale weight can obscure meaningful changes in body composition, such as fat loss accompanied by muscle gain, which may not be reflected in overall weight change [226].

Key Point: Body recomposition, i.e., losing fat while maintaining or gaining muscle, can dramatically improve body shape and health markers without significant changes in total body weight.

4.9.2 The Psychology of Scale Use

While many people have been conditioned to equate lower scale numbers with success, it is vital to recognize that the scale is just one tool among many for tracking progress. Daily body weight fluctuations can be substantial, sometimes amounting to several pounds in 24 hours due to changes in water, glycogen, and intestinal content. This variability can be discouraging for those trying to lose weight, leading some to believe that daily self-weighing is detrimental.

However, evidence suggests that daily self-weighing, when used as a tool for self-regulation rather than as a sole indicator of success, is associated with better weight loss and maintenance outcomes. People who weigh themselves daily are more likely to succeed at either losing weight or maintaining their weight after weight loss treatment, although it is also possible that these individuals are more motivated overall [226]. This is supported by several studies showing that daily self-weighing is associated with greater weight loss and maintenance [227–229].

The key is to use the scale as a feedback mechanism to inform healthy behaviors, not as the only measure of progress or self-worth. When combined with other indicators (such as body composition, measurements, performance, and subjective well-being) regular self-weighing can be a powerful tool for self-regulation and long-term success.

Key Point: The scale can be a useful tool for self-regulation when combined with other indicators of progress, but relying on it as the sole measure can obscure genuine improvements in health and body composition.

4.9.3 The Role of Muscle in Body Shape

A critical insight often overlooked in conventional weight loss approaches is that muscle tissue is the primary determinant of body shape and metabolic health. Without adequate muscle mass, the body lacks the structural framework that creates desired curves, definition, and proportions.

Muscle tissue serves multiple functions beyond aesthetics: it is metabolically active, burning calories at rest; it provides structural support for joints and bones; it improves insulin sensitivity and glucose metabolism; and it maintains functional capacity for daily activities. Rapid weight loss approaches that neglect muscle preservation often result in a "skinny fat" phenotype, i.e., individuals who achieve a lower scale weight but maintain high body fat percentages and poor metabolic health.

Furthermore, muscle tissue continues to develop and strengthen in response to appropriate stimuli (resistance training and adequate protein intake) throughout the lifespan, making body recomposition possible at any age. This contrasts sharply with the temporary nature of most scale-focused interventions, which typically result in weight regain within months or years.

Interestingly, the widely held belief that gradual, slower weight loss results in better long-term outcomes than rapid weight loss is not supported by current scientific evidence. As detailed by Casazza et al., this belief has been perpetuated for decades by health authorities, textbooks, and professional guidelines, largely as a reaction to the adverse effects observed with nutritionally insufficient very-low-calorie diets in the 1960s and the assumption that gradual lifestyle changes are more sustainable [226]. However, Casazza et al. review both observational and randomized controlled trial evidence and conclude that, in fact, the opposite is often true: greater initial weight loss is consistently associated with greater long-term success in maintaining weight loss, and rapid weight loss does not necessarily predispose individuals to greater weight regain [226]. This conclusion is supported by a comprehensive review by Astrup and Rossner, who found that rapid initial weight loss was beneficial for long-term maintenance [230], and by Nackers et al., who reported that participants with more rapid initial weight loss had better long-term outcomes than those who lost weight more slowly [231]. The authors explain that while the rationale for gradual weight loss is based on the idea that slow changes are more likely to be permanent, the empirical evidence does not support this; instead, those who lose more weight initially are more likely to maintain their weight loss over time. Importantly, however, it is critical that rapid initial weight loss is achieved through safe, nutritionally adequate, and sustainable methods, rather than through extreme calorie restriction or unhealthy practices. When approached in this way, a faster rate of weight loss can be associated with better long-term outcomes [231–233]. Ultimately, the focus should remain on the quality of weight lost (i.e., fat vs. muscle), ensuring that weight loss comes primarily from fat rather than muscle mass, and that health and sustainability are prioritized throughout the process.

> **Key Point:** Muscle tissue is essential for body shape, metabolic health, and functional capacity. Approaches that neglect muscle preservation often fail to achieve desired aesthetic and health outcomes.

4.9.4   Measuring Progress Beyond the Scale

Given the limitations of scale weight as a progress indicator, successful body recomposition requires alternative assessment methods that capture changes in body composition, function, and health. These include progress photographs taken under consistent lighting and positioning, which can reveal changes in body shape that scale weight cannot detect; body measurements of key areas such as waist, hips, arms, and thighs; clothing fit as a practical indicator of body shape changes; and performance metrics such as strength gains, endurance improvements, or functional movement assessments.

For those with access to more sophisticated tools, body composition analysis can offer a much more detailed look at changes in fat and muscle. One such method is the DEXA scan, which stands for Dual-Energy X-ray Absorptiometry. A DEXA scan is a non-invasive test that uses very low-dose X-rays to measure the amounts of fat, muscle, and bone within your body. It provides a precise breakdown of your body composition, including where fat and muscle are distributed. In many fitness centers, sports medicine clinics, and some private health and wellness practices, DEXA scans are marketed and routinely used to assess body fat percentage, visceral fat (fat around the organs), and lean muscle mass as part of body composition analysis. People who are on weight loss journeys, athletes, bodybuilders, and those interested in tracking their fitness progress often pay for DEXA scans solely for this purpose.

Other methods, such as bioelectrical impedance or hydrostatic (underwater) weighing, can also be used to estimate fat mass, lean mass, and bone density. Hydrostatic weighing, also known as underwater weighing, is a technique that determines body composition by measuring a person's weight while submerged in water compared to their weight on land. Because fat tissue is less dense than water and muscle tissue is denser, this method allows for an accurate estimation of body fat percentage based on how much water is displaced and the difference in weights. Hydrostatic weighing has long been considered a gold standard for body composition analysis in research settings. However, these techniques require specialized equipment and consistent protocols.

Most importantly, subjective measures of well-being (including energy levels, sleep quality, mood, and confidence) often improve dramatically during successful body recomposition, even when scale weight remains stable. These quality-of-life indicators are arguably more important than any numerical measurement.

As mentioned above, it is important to recognize that daily self-weighing, when combined with other behavioral strategies, can be a useful tool for weight management, but should not be the sole measure of progress. Studies show that people who weigh themselves daily lose more weight, maintain the lost weight, and are better able to prevent gaining or regaining weight than those who do not weigh themselves daily [227–229]. However, daily fluctuations should not be over-interpreted, as they often reflect changes in water and glycogen rather than true changes in fat or muscle mass [226].

> **Key Point:** Progress in body recomposition is best assessed through multiple indicators including photos, measurements, performance, and subjective well-being, rather than scale weight alone.

4.9.5   Nutritional and Training Implications

Successful body recomposition requires a fundamentally different approach than traditional weight loss. While weight loss can be achieved through caloric restriction alone, body recomposition is best supported by adequate protein intake and resistance training to promote muscle protein synthesis and maintenance.

Earlier research, such as Mori (2014), suggested that the timing of protein and carbohydrate intake after resistance exercise is particularly important for trained individuals, with immediate post-exercise consumption associated with greater nitrogen balance compared to delayed intake, while untrained individuals were less sensitive to timing [234]. However, other studies have found that if total daily protein intake is sufficient, the timing of protein intake after resistance exercise has minimal or no significant effect on muscle protein synthesis or nitrogen balance in both trained and untrained individuals [235–238].

The current consensus in the literature is that the so-called "anabolic window" after resistance training is much broader than previously thought, and that meeting daily protein needs is the most important factor for supporting muscle adaptation and body recomposition [235–238]. This means that individuals should focus on adequate total protein intake, paired with regular resistance training, rather than stressing about the precise timing of protein consumption. For most people, spacing protein intake evenly across meals and snacks throughout the day is a practical strategy to optimize muscle protein synthesis and support recovery.

Nutritional approaches should still emphasize nutrient-dense food choices and individualized meal planning, ideally with the advice of a registered dietitian. Unlike crash diets that severely restrict calories and often eliminate entire food groups, effective body recomposition protocols involve moderate caloric deficits, adequate protein, and balanced macronutrient intake to support both training and recovery.

> **Key Point:** Body recomposition requires adequate protein intake, resistance training, moderate caloric deficits, and patience. The timing of protein intake after exercise is much less important than once believed; what matters most is meeting your total daily protein needs and consistently engaging in resistance training [235–238].

4.9.6   What (Not) To Do: Recommendations

○ **Do not use scale weight as your only progress indicator.** Focus on body composition changes, measurements, photos, and how you feel rather than daily weight fluctuations.

○ **Do prioritize muscle preservation and growth.** Include adequate protein and resistance training in your routine.

○ **Do not pursue extreme caloric deficits.** Moderate deficits allow for fat loss while preserving muscle mass.

○ **Be patient with the process.** Body recomposition occurs more slowly than rapid weight loss but produces more sustainable and aesthetically pleasing results.

○ **Do not avoid resistance training for fear of weight gain.** Muscle gain is essential for achieving the body shape most people desire.

○ **Do track multiple progress indicators.** Use progress photos, measurements, performance metrics, and subjective well-being alongside occasional weigh-ins.

○ **Do not let daily weight fluctuations affect your mood or motivation.** Weight can vary 2-5 pounds daily due to hydration, hormones, and digestive contents.

○ **Do focus on meeting your daily protein needs.** While the timing of protein intake is less critical, consistent and sufficient protein intake throughout the day supports muscle maintenance and growth.

○ **Do consider individualized nutrition planning.** Working with a registered dietitian can help ensure your calorie and nutrient targets are tailored to your needs and support both training and recovery.

Main Takeaway: The number on the scale is not the ultimate measure of health, progress, or attractiveness. Focus on changing your body composition by losing fat and maintaining or gaining muscle, rather than simply losing weight. This approach leads to better aesthetic outcomes, improved health markers, and more sustainable results.

4.10   Aerobic vs. Anaerobic Exercise

A central question in exercise and weight management is whether aerobic (cardiovascular) or anaerobic (resistance/strength) training is more effective for fat loss and improving body composition. This section reviews the latest evidence from randomized controlled trials, systematic reviews, and narrative syntheses, with a focus on adults with overweight or obesity.

4.10.1   Cardio vs. Resistance Training

Multiple high-quality systematic reviews and meta-analyses have consistently found that both aerobic and resistance training can lead to significant improvements in body composition, but their effects differ in important ways:

○ **Aerobic training (AT)** is generally more effective than resistance training (RT) for reducing total body weight, fat mass, and visceral (abdominal) fat, especially when energy expenditure is matched [239–242].

o **Resistance training (RT)** is superior for preserving or increasing lean muscle mass during weight loss, and is particularly effective for reducing waist-to-hip ratio and abdominal fat, especially in men [239, 241, 242].

o **Combined training (CT)**, i.e., programs incorporating both aerobic and resistance modalities, consistently produces the greatest improvements in body composition, maximizing fat loss while minimizing loss of muscle mass [239, 241–243].

Beyond body composition, aerobic and resistance training have distinct and sometimes complementary effects on cardiovascular health markers, particularly lipid profiles. Aerobic exercise is well-established to increase high-density lipoprotein cholesterol (HDL), often referred to as "good" cholesterol because higher levels are associated with a lower risk of heart disease, and, to a lesser extent, reduce low-density lipoprotein cholesterol (LDL), commonly known as "bad" cholesterol due to its role in promoting atherosclerosis and cardiovascular risk [105–107]. Resistance training may also improve lipid profiles, but the effects are generally smaller and more variable [106, 107].

4.10.2   Practical Implications

A recent umbrella review of 12 systematic reviews (149 studies) found that exercise interventions in adults with overweight or obesity led to average weight losses of 1.5 to 3.5 kg, fat losses of 1.3 to 2.6 kg, and clear reductions in visceral fat (i.e., the type of fat stored around internal organs) even when the total amount lost wasn't always large. These changes were seen over intervention periods typically lasting 2 to 12 months [239]. No significant differences in fat loss were found between aerobic exercise and high-intensity interval training (HIIT) when people expended the same amount of energy. However, resistance training alone did not significantly reduce visceral fat, but was effective in helping people preserve lean muscle mass during weight loss (on average, maintaining 0.8 kg more muscle) [239].

A large interventional study in middle-aged adults with obesity found that all exercise modalities (aerobic, HIIT, and resistance) led to significant reductions in body weight, waist-to-hip ratio, and body fat percentage compared to diet alone, with the greatest reductions in body fat percentage observed in the HIIT group for women and in the resistance group for men [242]. Importantly, all exercise groups maintained muscle mass, while diet alone did not produce significant changes in body composition.

Narrative reviews and network meta-analyses confirm these findings: aerobic training is most effective for fat loss, resistance training is best for muscle preservation, and the combination yields optimal results for both outcomes [239–241, 243].

4.10.3 Key Mechanisms and Sex Differences

Aerobic exercise promotes fat loss primarily through increased energy expenditure and enhanced fat oxidation, especially at moderate to high intensities. Resistance training increases muscle mass and resting metabolic rate, which supports long-term weight management and improves metabolic health [241, 242]. Resistance training is particularly effective for reducing abdominal fat and waist-to-hip ratio in men, likely due to greater muscle mass and fat oxidation capacity [242].

The effects of exercise on lipid metabolism are also influenced by individual baseline characteristics. The HERITAGE Family Study by Couillard et al. (2001) demonstrated that the HDL-raising effect of endurance exercise is most pronounced in men with both low HDL cholesterol at baseline and greater abdominal obesity [106]. In this group, 20 weeks of supervised endurance training led to a significant increase in HDL, with the increase closely correlated with reductions in abdominal subcutaneous fat. In contrast, men with isolated low HDL but without excess abdominal fat did not experience significant HDL increases from exercise.

Recent studies highlight the effectiveness of hybrid online/offline exercise programs for improving adherence and engagement, especially in the context of public health challenges such as COVID-19. These models are scalable and can deliver significant improvements in body composition and metabolic health [242].

4.10.4 Cardiovascular Health and Lipid Metabolism

The cardiovascular benefits of exercise extend beyond weight loss and body composition. Muscella et al. (2020) reviewed the effects of exercise training (particularly endurance (aerobic) training) on lipid metabolism and coronary heart disease risk. They found that aerobic exercise consistently lowers total cholesterol and LDL, increases HDL, and improves the maturation and function of HDL particles. These changes enhance reverse cholesterol transport, a key anti-atherosclerotic mechanism, and reduce the risk of coronary heart disease. Importantly, these benefits are observed even in the absence of significant changes in body composition, underscoring the independent value of regular exercise for cardiovascular health [107]. These findings are supported by other recent reviews and clinical trials, which confirm that aerobic exercise favorably alters lipid profiles and reduces cardiovascular risk, regardless of weight loss [105–107].

4.10.5   Limitations and Considerations

While both aerobic and resistance training are effective, the magnitude of weight loss from exercise alone is modest compared to interventions that combine exercise with dietary changes. Exercise is critical for preserving muscle mass, improving metabolic health, and supporting long-term weight maintenance, but should ideally be paired with caloric control for maximal fat loss [239, 241, 243].

It is also important to recognize that the lipid-modifying effects of exercise are not uniform across all individuals. As shown by Couillard et al. (2001), baseline HDL levels and abdominal fat can influence the degree of improvement in lipid profiles [106]. Similarly, Kodama et al. (2007) found that the duration of each exercise session is a key determinant of HDL response, and that individuals with higher cholesterol and lower BMI benefit most [105].

> **Key Point:** Aerobic training is most effective for fat loss and improving lipid profiles, resistance training is best for muscle preservation, and combining both yields the greatest improvements in body composition and cardiovascular health. Exercise alone produces modest weight loss, but is essential for maintaining muscle mass, improving lipid metabolism, and reducing cardiovascular risk during weight loss.

4.10.6   What (Not) To Do: Recommendations

○ **Do not rely on resistance training alone for maximal fat loss.** While RT preserves muscle, aerobic or combined training is needed for optimal fat reduction and cardiovascular benefit.

o **Do not neglect resistance training during weight loss.** Preserving muscle mass is critical for health, function, and long-term weight maintenance.

o **Do combine aerobic and resistance training.** This approach maximizes fat loss, preserves muscle, improves lipid profiles, and enhances metabolic and cardiovascular health.

o **Do not expect large weight losses from exercise alone.** For substantial fat loss, combine exercise with dietary caloric control.

o **Be mindful of individual preferences and limitations.** Choose exercise modalities that are enjoyable and sustainable for long-term adherence, and consider baseline lipid profiles and body composition when tailoring exercise prescriptions.

Main Takeaway: For optimal fat loss, body composition, and cardiovascular health, combine aerobic and resistance training, ideally alongside dietary changes. Aerobic exercise is best for fat loss and improving lipid profiles, resistance training for muscle preservation, and the combination for overall health and sustainable results.

4.11   Aging and Muscle Loss

Aging brings a gradual but significant decline in muscle mass and strength, a process known as sarcopenia. This age-related muscle loss can profoundly affect health, functional ability, and quality of life, yet it often goes unnoticed until its consequences become severe. Understanding why and when muscle loss occurs, as well as how to counteract it, is vital for lifelong health and independence.

4.11.1   What Is Sarcopenia?

Sarcopenia refers to the progressive loss of skeletal muscle mass and function that occurs with advancing age. The concept was first formalized in the late 1980s to describe the loss of muscle mass with aging, but definitions have since evolved to include not only muscle mass but also muscle strength and physical performance [244, 245]. The most widely accepted modern definitions, such as those from the European Working Group on Sarcopenia in Older People (EWGSOP2), require the presence of low muscle strength as the primary criterion, with confirmation by low muscle quantity or quality, and further classification of severity based on physical performance [245].

The decline in muscle mass is driven by a combination of factors, including hormonal changes (such as reductions in testosterone, growth hormone, and IGF-1), decreased physical activity, reduced protein intake and synthesis, chronic inflammation, and the presence of comorbid diseases. Importantly, recent research has shown that the loss of muscle strength (sometimes called dynapenia) often exceeds the loss of muscle mass, indicating that changes in muscle quality and neuromuscular function are also critical [244, 245].

> **Key Point:** Muscle loss with aging is driven by hormonal changes, reduced physical activity, inadequate protein intake, and chronic disease, but it is not inevitable and can be slowed or reversed with proper interventions.

4.11.2　When Does Muscle Loss Start?

The onset and rate of muscle loss with age have been quantified in both cross-sectional and longitudinal studies. Cross-sectional studies comparing young adults (typically aged 18–30) to older adults (over 70) have found that muscle mass declines by approximately 4.7% per decade in men and 3.7% per decade in women, with the rate of loss accelerating after age 70 [244]. Longitudinal studies, which follow the same individuals over time, provide more precise estimates: in people aged 75, muscle mass is lost at a rate of about 0.8–1.0% per year in men and 0.6–0.7% per year in women. By age 80, up to 50% of men and women may meet criteria for sarcopenia, depending on the definition and population studied [244].

However, the decline in muscle strength is even more pronounced. Longitudinal data show that at age 75, strength is lost at a rate of 3–4% per year in men and 2.5–3% per year in women, meaning that strength declines two to five times faster than muscle mass [244]. This dissociation is important because low muscle strength is a more consistent predictor of disability, hospitalization, and mortality than low muscle mass alone [244, 245]. For example, studies have shown that individuals in the lowest quartile of handgrip strength have nearly double the risk of developing disability or dying compared to those in the highest quartile, even after adjusting for muscle mass and other factors [244].

The consequences of sarcopenia are far-reaching. Reduced muscle mass and strength lead to impaired mobility, increased risk of falls and fractures, loss of independence, and higher healthcare costs. Sarcopenia is also associated with metabolic dysfunction, including insulin resistance and increased risk of type 2 diabetes, as well as poorer outcomes in chronic diseases such as heart failure and chronic obstructive pulmonary disease [245].

> **Key Point:** Muscle loss accelerates with age, especially after 50, contributing to frailty, metabolic disease, and loss of independence. Prevention is key.

4.11.3   How to Maintain Muscle Mass Through Aging

A large body of evidence demonstrates that sarcopenia is not an inevitable consequence of aging and can be significantly slowed or even reversed with appropriate interventions. The most effective strategies are regular resistance exercise and adequate protein intake. A comprehensive meta-analysis by Peterson et al. synthesized data from 47 studies (over 1,000 participants aged 50–92) and found that resistance exercise leads to substantial improvements in muscle strength in older adults [246]. The studies included a range of training protocols, with durations from 6 to 52 weeks, intensities from 40% to 85% of one-repetition maximum (1RM), and frequencies of 1–3 sessions per week. Across all studies, resistance exercise increased strength by 24–33% depending on the muscle group, with absolute gains ranging from 9.8 to 31.6 kg. Notably, higher-intensity training ($\geq$70% 1RM) was associated with greater improvements in strength. These gains were observed in both men and women, and across a wide age range, including participants in their 80s [246].

Other systematic reviews and randomized controlled trials confirm that resistance training not only increases muscle strength but also improves muscle mass, physical performance (such as gait speed and chair rise ability), and quality of life in older adults [244, 245]. Even short-term interventions (8–12 weeks) can produce meaningful gains, and longer-term programs yield even greater benefits. Importantly, resistance training is effective even in very old adults and those with chronic diseases, although the magnitude of response may be somewhat blunted in the oldest old (those aged 85 years and older) or those with severe frailty [244].

Adequate protein intake is also essential for maintaining muscle mass. Observational studies show that many older adults consume less than the recommended 0.8 g/kg/day of protein, and intakes of 1.2–1.5 g/kg/day may be optimal for preserving muscle in older age [244, 245]. Protein supplementation, especially when combined with resistance training, has been shown to increase lean mass and improve functional outcomes in older adults with sarcopenia [244].

Other factors that contribute to muscle maintenance include management of chronic diseases, reduction of inflammation, correction of vitamin D deficiency, and maintenance of physical activity beyond resistance training (such as walking, balance, and flexibility exercises) [244, 245]. Testosterone therapy has also been investigated as a potential intervention for muscle loss in older adults. Studies suggest that testosterone administration can increase muscle mass, improve muscle strength, and enhance physical function, particularly in older men with low or low-normal testosterone levels [247–249]. For example, combining testosterone supplementation with resistance training, vitamin D, calcium, and protein has been shown to improve muscle strength and quality of life in men aged 70 years and older [247]. However, the benefits may vary between individuals, and potential side effects (such as increased hematocrit and cardiovascular risk) should be carefully considered when evaluating testosterone therapy for sarcopenia or age-related muscle loss [250].

> **Key Point:** Muscle loss with age is not inevitable; regular resistance training and sufficient protein intake can maintain and even increase muscle mass and strength, preserving health and independence.

4.11.4 What (Not) To Do: Recommendations

o **Do not rely solely on aerobic exercise.** While important for cardiovascular health, aerobic activity alone does little to preserve or build muscle mass. Resistance training is essential for maintaining strength and function [245, 246].

o **Do not accept muscle loss as an unchangeable part of aging.** High-quality evidence shows that older adults can build muscle and strength with proper training and nutrition, even into their 80s and beyond [244, 246].

○ **Do prioritize resistance training and adequate protein.** These interventions are the most effective for preventing and reversing sarcopenia. Aim for at least two sessions per week of resistance exercise, and consume 1.2–1.5 g/kg/day of protein if possible [244, 245].

○ **Be proactive.** Start building and maintaining muscle mass early (ideally in your 30s and 40s) to have a higher "reserve" as you age. However, it is never too late to benefit from starting resistance training [244].

○ **Do not ignore changes in strength, balance, or function.** Early intervention is more effective than trying to regain lost muscle after significant decline. Regular assessment of muscle strength (e.g., grip strength or chair stand tests) can help identify those at risk [245].

Main Takeaway: Age-related muscle loss begins as early as your 30s but accelerates after 50. Proactive resistance training and protein intake can substantially slow or reverse this process, supporting vitality, metabolic health, and independence for life.

4.12   The DASH and Mediterranean Diets

Dietary patterns such as the DASH (Dietary Approaches to Stop Hypertension) and Mediterranean diets have become widely recommended for preventing and managing high blood pressure. But what does the best current research say about how well these diets work, who benefits most, and why? Here we review the latest systematic reviews, meta-analyses, and clinical trials, highlighting not just the results but the methods used to reach these conclusions.

To answer whether the DASH and Mediterranean diets lower blood pressure, researchers have used several rigorous approaches. Randomized controlled trials assign people to follow a specific diet (like DASH) or a control diet, then directly measure blood pressure and other health outcomes before and after the intervention. Meta-analyses combine the results of many such trials, providing a more precise estimate of the overall effect, and can explore whether diet effects differ by age, sodium intake, or other factors. Systematic reviews look at all available evidence, assessing the quality of the studies and the consistency of the findings. Observational cohort studies follow large groups of people over time to see if those who eat more like the DASH or Mediterranean pattern have lower rates of developing hypertension. Newer studies also use metabolomics (analyzing blood or urine samples for thousands of small molecules) to objectively track what people are eating and how it affects their bodies.

> **Key Point:** The effects of the DASH and Mediterranean diets on blood pressure have been studied using randomized controlled trials, meta-analyses, systematic reviews, large cohort studies, and advanced metabolomics, providing a robust and nuanced evidence base.

4.12.1   DASH Diet: Methods and Results

A major systematic review and meta-analysis pooled 30 randomized controlled trials (involving 5,545 participants) to assess the blood pressure-lowering effects of the DASH diet in adults, both with and without hypertension [251]. The researchers included only high-quality randomized control trials that compared the DASH diet to a control diet, regardless of whether participants were taking blood pressure medications, had other health conditions, or were following other lifestyle interventions (like sodium restriction or weight loss). Both random-effects and fixed-effects statistical models were used to calculate the average difference in blood pressure during follow-up between the DASH and control groups. The analysis also included multiple subgroup and sensitivity analyses (such as by age, baseline blood pressure, sodium intake, and study quality) to see if the results held up in different populations.

The results were clear and robust: across all studies, the DASH diet reduced systolic blood pressure by an average of 3.2 mmHg and diastolic by 2.5 mmHg, both highly statistically significant (P < 0.001). While these numbers may seem small, even modest reductions in blood pressure at the population level are associated with meaningful reductions in the risk of heart attack and stroke. For example, a 2 mmHg reduction in systolic blood pressure is estimated to reduce the risk of cardiovascular events by about 7–10% [251].

Importantly, the effect was similar whether or not people were already diagnosed with hypertension or were taking antihypertensive medications. The blood pressure reduction was greater in studies where average sodium intake was higher (>2400 mg/day), and more pronounced in younger participants (under age 50). The findings remained consistent when looking only at studies with rigorous designs, intention-to-treat analyses, or office blood pressure measurements. No single study drove the results, and no evidence of publication bias was found.

In addition, a recent systematic review and meta-analysis of 12 large cohort and cross-sectional studies (over 115,000 adults) found that people with the highest adherence to the DASH diet had a 19% lower risk of developing hypertension compared to those with the lowest adherence [252]. The results were consistent whether hypertension was diagnosed by direct blood pressure measurement or by self-report. Sensitivity analyses supported the robustness of these findings.

> **Key Point:** The DASH diet consistently lowers blood pressure in both hypertensive and normotensive adults, and high adherence reduces the risk of developing hypertension. The effect is robust across study designs, populations, and methods of blood pressure assessment.

4.12.2   Mediterranean Diet: Methods and Results

The Mediterranean diet has also been studied in large randomized control trials and meta-analyses, though its effects on blood pressure appear somewhat smaller than those of the DASH diet. Systematic reviews and meta-analyses of randomized control trials have found that, compared to usual or low-fat diets, the Mediterranean diet lowers systolic blood pressure by about 1.5–3 mmHg and diastolic by 0.9–1.6 mmHg [253]. While these effects are statistically significant and may contribute to reduced cardiovascular risk, they are consistently smaller than the reductions seen with the DASH diet. The Mediterranean diet's benefits are likely due to its high content of unsaturated fats (especially olive oil), fruits, vegetables, whole grains, nuts, and fish, as well as moderate alcohol (wine) intake.

> **Key Point:** The Mediterranean diet lowers blood pressure, but the effect is smaller than that of the DASH diet. Its benefits are likely due to a combination of healthy fats, plant foods, and overall dietary pattern.

Both the DASH and Mediterranean diets are naturally lower in sodium than typical Western diets, but the effect of additional salt reduction has been studied separately and in combination with these dietary patterns. Meta-analyses of randomized control trials show that reducing sodium intake by about 2.3 grams per day lowers systolic blood pressure by 4–8 mmHg in people with hypertension, but only 1–2 mmHg in those with normal blood pressure [253]. Interestingly, when a low-sodium DASH diet is compared to a DASH diet with higher sodium, the blood pressure-lowering effect of salt reduction is less than additive; meaning that combining the two does not double the effect. For example, in the DASH-Sodium trial, reducing both sodium and following the DASH diet together lowered systolic blood pressure by 8.9 mmHg, compared to 6.7 mmHg for salt reduction alone, and 3.0 mmHg for DASH alone [253]. The overlap is likely due to both interventions acting on similar physiological pathways.

> **Key Point:** Reducing salt intake lowers blood pressure, espe-
> cially in people with hypertension. Combining salt reduction
> with the DASH diet produces the greatest blood pressure drop,
> but the effects are not fully additive.

4.12.3   How Do These Diets Work?

The DASH and Mediterranean diets lower blood pressure through a combination of high intakes of potassium, magnesium, calcium, fiber, and antioxidants, while reducing sodium and saturated fat. These nutrients help relax blood vessels, promote sodium excretion, and reduce inflammation and oxidative stress. The DASH diet in particular acts like a mild diuretic, helping the kidneys eliminate sodium and water, thereby reducing blood pressure.

Recent metabolomics studies have taken this a step further by directly measuring thousands of compounds in urine or blood samples of people following these diets [254, 255]. In a controlled feeding study, researchers identified hundreds of unique compounds from DASH-style foods in urine (called food-specific compounds, or FSCs), confirming with high precision that participants were adhering to the diet. Although no single unmetabolized food compound was directly linked to blood pressure levels, several endogenous and food-related metabolites were associated with changes in blood pressure, suggesting complex metabolic pathways at work. These methods open the door to more objective monitoring of dietary adherence and to understanding why certain diets work better for some individuals.

Emerging evidence suggests that the DASH diet may also benefit people with heart failure, improving blood pressure, arterial compliance, cardiac function, and quality of life [255]. However, most studies in heart failure are small and non-randomized, and more research is needed for firm recommendations. The use of metabolomics and personalized nutrition approaches holds promise for tailoring these diets to individual needs and optimizing their benefits.

Despite strong evidence, real-world adherence to the DASH or Mediterranean diets remains low. Barriers include food preferences, cultural habits, cost, and access to fresh food. Behavioral strategies, self-monitoring, and tailored interventions can help improve adherence. Objective biomarkers from metabolomics may soon allow clinicians to track dietary adherence more accurately.

> **Key Point:** The DASH and Mediterranean diets lower blood pressure through multiple nutrients and bioactive compounds acting synergistically. Metabolomics can objectively track adherence and metabolic effects, and may help personalize dietary recommendations in the future.

4.12.4   What (Not) To Do: Recommendations

o **Do not wait for high blood pressure to start a healthy diet.** The DASH and Mediterranean diets are effective for prevention as well as treatment.

o **Do emphasize whole foods rich in potassium, calcium, magnesium, and fiber:** fruits, vegetables, whole grains, legumes, nuts, and low-fat dairy.

o **Do limit processed foods and added salt.** Most dietary sodium comes from packaged and restaurant foods.

o **Do not rely on supplements alone.** The synergy of nutrients in whole foods is key to these diets' benefits.

o **Do tailor dietary advice to individual preferences, culture, and barriers to adherence.**

o **Do use objective tools (e.g., urine metabolite tests) to monitor progress when possible.**

> **Main Takeaway:** The DASH diet (and, to a lesser extent, the Mediterranean diet) are proven, practical, and safe strategies

for lowering blood pressure and preventing hypertension. Early adoption, a focus on whole foods, and individualized support maximize their effectiveness.

4.13   Sodium Intake

Sodium intake is a central factor in the development and management of high blood pressure (hypertension). Excessive dietary sodium, primarily from salt (sodium chloride), is a well-established contributor to elevated blood pressure and increased cardiovascular risk. This section reviews the evidence on the effects of salt reduction, compares it with the impact of dietary patterns such as the DASH and Mediterranean diets (see Section 4.12), and discusses their interplay and practical implications for blood pressure control.

4.13.1   How Does Sodium Affect Blood Pressure?

Sodium is essential for maintaining fluid balance and nerve function, but most people consume far more than the physiological requirement. The kidneys are responsible for excreting excess sodium, but when intake is chronically high, the body retains more water, leading to increased blood volume and, consequently, higher blood pressure. This effect is particularly pronounced in individuals who are "salt-sensitive," a group that includes many people with hypertension, older adults, those with chronic kidney disease, and certain ethnic groups. High sodium intake also activates hormonal and neural pathways that further raise blood pressure and can contribute to vascular stiffness and endothelial dysfunction [253].

> **Key Point:** High sodium intake increases blood pressure through fluid retention and activation of hormonal and vascular pathways, especially in salt-sensitive individuals.

4.13.2   How Much Should Sodium Intake Be?

Major health organizations, including the World Health Organization and leading hypertension guidelines, recommend reducing sodium intake to less than 2,000 mg per day (about 5 grams of salt) for adults, and ideally closer to 1,500 mg per day for those with high blood pressure [253]. Meta-analyses of randomized controlled trials show that reducing sodium intake by about 2.3 grams per day lowers systolic blood pressure by 4–8 mmHg in people with hypertension, and by 1–2 mmHg in those with normal blood pressure. The greater the reduction in sodium, the larger the drop in blood pressure, with the most pronounced effects seen in those with higher baseline sodium intake and those who are salt-sensitive. Importantly, even modest, sustained reductions in sodium intake across the population can have significant public health benefits by reducing the incidence of heart attack, stroke, and kidney disease [253].

> **Key Point:** Reducing sodium intake is one of the most effective single dietary strategies for lowering blood pressure, especially in people with hypertension or high baseline sodium intake.

4.13.3   Salt Reduction vs. DASH

The DASH and Mediterranean diets are both recommended for blood pressure control and cardiovascular health. These dietary patterns emphasize fruits, vegetables, whole grains, nuts, legumes, and low-fat dairy, while limiting red meat, processed foods, and added sugars. Both diets are naturally lower in sodium than typical Western diets, but their benefits extend beyond salt reduction, as they are rich in potassium, magnesium, calcium, fiber, and antioxidants, all of which contribute to blood pressure regulation [251–253].

A pivotal study, the DASH-Sodium trial, directly compared the effects of sodium reduction, the DASH diet, and their combination. The results showed that reducing sodium intake alone, while eating a typical diet, lowered systolic blood pressure by 6.7 mmHg. Adopting the DASH diet alone (without additional sodium restriction) lowered systolic blood pressure by 3.0 mmHg. However, combining the DASH diet with low sodium intake produced the largest reduction: an average drop of 8.9 mmHg in systolic blood pressure [253]. This pattern was even more pronounced in people with hypertension, where the combined intervention led to reductions of up to 20 mmHg in those with very high baseline blood pressure.

Interestingly, the effects of salt reduction and the DASH diet are not fully additive. That is, the combined effect is less than the sum of each intervention separately, likely because both approaches act on overlapping physiological pathways. The Mediterranean diet, while also beneficial, generally produces smaller reductions in blood pressure compared to the DASH diet or salt reduction alone [253].

> **Key Point:** Salt reduction alone produces a larger drop in blood pressure than adopting the DASH or Mediterranean diet alone, but the combination of salt reduction and a healthy dietary pattern yields the greatest benefit.

4.13.4   Mechanisms and Practical Implications

Salt reduction lowers blood pressure primarily by decreasing fluid retention and blood volume, while the DASH and Mediterranean diets provide additional benefits through increased intake of potassium, magnesium, calcium, fiber, and antioxidants. These nutrients help relax blood vessels, promote sodium excretion, and reduce inflammation and oxidative stress. The DASH diet, in particular, acts like a mild diuretic, enhancing the body's ability to eliminate sodium and water [251, 252].

From a practical standpoint, reducing salt intake is a focused and achievable first step for most people, as it often involves simple changes such as cooking more at home, using less salt in recipes, choosing lower-sodium packaged foods, and being mindful of restaurant meals. Once salt reduction becomes a habit, adopting a broader dietary pattern like DASH or Mediterranean can provide further benefits for blood pressure and overall health. Notably, these diets are naturally lower in sodium, so adopting them often leads to additional reductions in salt intake even without strict tracking.

> **Key Point:** For blood pressure control, start by reducing salt intake, then adopt a healthy dietary pattern for maximal benefit. The combination of both strategies yields the greatest reduction in blood pressure and risk of cardiovascular disease.

4.13.5   What (Not) To Do: Recommendations

○ **Do not assume that adopting a healthy diet alone is enough if salt intake remains high.** Salt reduction is a powerful and necessary step, especially for those with hypertension.

○ **Do start by reducing sodium intake.** Aim for less than 2,300 mg per day (about one teaspoon of salt), and ideally closer to 1,500 mg per day if you have high blood pressure.

○ **Do adopt the DASH or Mediterranean diet for additional benefits.** These patterns not only support blood pressure control but also improve overall cardiovascular and metabolic health.

○ **Do not rely solely on processed or restaurant foods, which are major sources of hidden sodium.** Prepare meals at home when possible, use herbs and spices for flavor, and read labels carefully.

○ **Do not expect the effects of salt reduction and dietary pattern to be fully additive.** Both work through similar mechanisms, so the combined effect, while greater, is less than the sum of each alone.

○ **Do personalize your approach.** Some people may find it easier to start with salt reduction, while others may prefer to focus on overall dietary quality. Both strategies are valid and can be combined for maximal benefit.

> **Main Takeaway:** For lowering blood pressure, reducing salt intake is the most effective single dietary step, but combining salt reduction with the DASH or Mediterranean diet provides the greatest overall benefit. Start by cutting back on sodium, then adopt a healthy dietary pattern for optimal results.

4.14 Brown Rice vs. White Rice

Brown rice is bad for you because it is "full of phytic acid and anti-nutrients, contains arsenic, and will harm your gut". In contrast, white rice is supposed to be superior for health.

4.14.1 Scientific Analysis and Rebuttal

This claim is not supported by the best available scientific evidence. While it is true that brown rice contains more phytic acid (an "anti-nutrient") and may have higher arsenic content than white rice, these concerns are often exaggerated and must be weighed against the broader nutritional and health context. Furthermore, the assertion that brown rice is uniquely harmful to the gut is contradicted by direct evidence from human studies.

Brown rice contains more arsenic than white rice because arsenic accumulates in the outer bran layer of the grain. During the milling process that transforms brown rice into white rice, this bran layer (along with the germ) is removed, resulting in white rice that is lower in arsenic content. This is because white rice is essentially the starchy endosperm that remains after the bran and germ are polished away, a process which also strips away much of the fiber and micronutrients present in brown rice

However, the absolute risk from dietary arsenic in rice is low for most people, especially when rice is washed thoroughly and cooked in excess water, which can reduce arsenic content by up to 50% [256, 257]. Many other common foods (such as seafood, mushrooms, apples, and grapes) also contain trace amounts of arsenic. Dietary guidelines recommend moderation and variety rather than avoidance. Phytic acid, while it can reduce the absorption of some minerals, also acts as an antioxidant and may have health benefits [258]. For most people eating a varied diet, the impact of phytic acid is negligible.

4.14.2   Gut Health and the Microbiota

The claim that brown rice is harmful to gut health is directly contradicted by evidence from the GENKI study, a nested cohort investigation conducted in Japan. This study specifically examined the gut microbiota of habitual brown rice eaters compared to those who consumed white rice [259]. The methodology was rigorous: 109 healthy adults (18 males, 91 females, average age approximately 54 years) provided stool samples for metagenomic analysis, and their dietary habits were assessed via detailed questionnaires. The study employed next-generation DNA sequencing to characterize the gut microbiota at the phylum, genus, and species levels. Statistical analysis included principal component analysis, correlation matrices, and both parametric and non-parametric tests to ensure robust results.

The study found that brown rice eaters had a gut microbiota profile characterized by higher levels of butyrate-producing bacteria compared to white rice eaters. The study concluded that brown rice consumption supports a gut microbiota profile rich in butyrate producers, which are linked to gut health, immune regulation, and anti-inflammatory effects. Butyrate is a short-chain fatty acid that serves as an energy source for colon cells and helps maintain gut barrier function.

In contrast, brown rice eaters had significantly lower levels of potentially harmful bacteria (0.018% in brown rice eaters vs. 1.6% in white rice eaters). The prevalence of other beneficial genera was also high in brown rice eaters. Importantly, these differences were independent of other dietary factors; correlation analysis showed that the changes in microbiota were not explained by differences in intake of vegetables, legumes, or other foods. The authors noted that the influence of brown rice on the microbiota was substantial and likely beneficial, contradicting claims of harm [259].

> **Key Point:** Brown rice consumption is associated with a gut microbiota profile rich in butyrate-producing bacteria, which are linked to gut health and metabolic benefits [259].

4.14.3   Metabolic Health and Diabetes Risk

The most comprehensive evidence on rice and metabolic health comes from a recent systematic review and meta-analysis, which synthesized data from 19 studies (8 prospective cohorts and 11 randomized controlled trials) including over 700,000 participants [260]. They included prospective cohort studies with at least one year of follow-up and randomized controlled trials with a minimum two-week intervention, focusing on healthy adults without diabetes at baseline. Only studies that specified brown or white rice as the exposure or intervention were included, and all had to report multivariable-adjusted relative risks for type 2 diabetes (T2D) or changes in cardiometabolic biomarkers.

The review found that high intake of white rice was associated with a 16% increased risk of T2D when comparing the highest to lowest intake categories. A threshold effect was observed: above 300 g/day (cooked), each additional 158 g/day serving of white rice was associated with a 13% higher risk of T2D. In contrast, brown rice intake was associated with an 11% lower risk of T2D comparing highest to lowest intake, with a linear dose-response: each 50 g/day increment of brown rice was associated with a 13% lower risk of T2D. Randomized controlled trials found that substituting brown rice for white rice led to a modest increase in HDL ("good") cholesterol, but no significant differences in other cardiometabolic biomarkers (fasting glucose, HbA1c, LDL cholesterol, triglycerides, waist circumference). The overall strength of evidence was moderate for cohort studies and moderate to low for RCTs, mainly due to limited sample size and short duration of trials. The meta-analysis included studies from diverse regions, with most brown rice data coming from US cohorts. All studies adjusted for key confounders, and sensitivity analyses confirmed the robustness of the results [260].

Key Point: High white rice intake increases diabetes risk,
while brown rice is associated with lower risk. The difference is
likely due to higher fiber, micronutrients, and effects on the gut
microbiota [260].

4.14.4   Butyrate and Metabolic Effects

The beneficial effects of brown rice on gut microbiota may translate into
improved metabolic health via increased production of butyrate. This is
supported by a randomized, triple-blind, placebo-controlled trial in patients
with T2D, which investigated the effects of oral butyrate supplementation
(NaBut) for six weeks [261]. In this trial, 42 patients with T2D were
randomly allocated to receive either NaBut (n=21) or placebo (n=21). The
study was carefully designed: it was triple-blind, meaning that participants,
clinicians, and outcome assessors were all blinded to group allocation. Serum
concentrations of metabolic parameters, glutathione peroxidase, nitric oxide,
and blood pressure were assessed before and after the intervention.

The results showed that within the NaBut group, systolic and diastolic blood pressure were significantly reduced. Blood sugar two hours
postprandial also decreased in both groups, but there were no significant
between-group differences for this or for insulin resistance, lipid profile, or
other metabolic parameters. Notably, NaBut supplementation increased
total cholesterol and LDL cholesterol compared to baseline, but again, these
changes were not significant when compared to placebo. The trial was short
(six weeks) and small, so longer studies are needed to clarify the effects.
Nonetheless, these results suggest that butyrate, produced by gut bacteria
favored by brown rice consumption, may have beneficial effects on blood
pressure and metabolic health, though more research is needed [261].

Key Point: Butyrate, produced by gut bacteria favored by
brown rice, has beneficial effects on blood pressure and may
support metabolic health [261].

4.14.5   Summary: Claim vs. Evidence

The claim that brown rice is harmful due to anti-nutrients, arsenic, and
negative effects on gut health is not supported by scientific evidence. Instead,
brown rice is associated with a healthier gut microbiota (higher levels of
butyrate-producing bacteria and lower levels of potentially harmful bacteria),

and higher intake is linked to lower risk of type 2 diabetes. Randomized trials show modest benefits for HDL cholesterol and blood pressure. The type of rice consumed is less important than overall dietary pattern and context [259–261].

Concerns about arsenic and phytic acid are manageable with proper preparation and a varied diet. However, the amount of arsenic remaining in rice after thorough washing and cooking is well below established safety thresholds for food, and similar low levels of arsenic can also be found in other foods, such as apple juice [262]. Regulatory agencies like the FDA routinely monitor arsenic levels in foods, and their findings show that both rice and apple juice generally contain only low levels of inorganic arsenic, far below levels of concern for health in the context of a balanced diet.

> **Conclusion:** The popular claim that brown rice is "bad for you" is not supported by scientific evidence. Both brown and white rice can be part of a healthy diet, but replacing white rice with brown rice or other whole grains may offer modest metabolic and gut health benefits, especially for those at risk of diabetes.

4.14.6   What (Not) To Do: Recommendations

- **Do not fear brown rice due to anti-nutrients or arsenic.** Wash rice thoroughly and eat a varied diet to minimize risk.

- **Do consider including brown rice or other whole grains.** These are associated with better gut health and lower diabetes risk.

- **Do not assume white rice is inherently unhealthy.** The overall dietary pattern, portion size, and context matter most.

- **Do focus on minimally processed, fiber-rich foods.** The benefits of brown rice are largely due to its fiber and micronutrient content.

- **Be skeptical of extreme claims about single foods.** Nutrition science supports moderation, variety, and context over rigid rules.

> **Main Takeaway:** Brown rice is not "bad for you." It supports a healthier gut microbiota, is linked to lower diabetes risk, and is safe for most people when prepared properly. The choice between white and brown rice should be based on personal preference, cultural context, and overall dietary quality, not on fear or misinformation [259–261].

4.15 Arsenic and Disease Susceptibility

Arsenic is a naturally occurring element that contaminates groundwater in many parts of the world, leading to chronic exposure for tens of millions of people. Regions at highest risk include Bangladesh, India, China, and parts of the Americas, with significant numbers also affected in Europe and the US [263, 264]. In the US, around 12% of public water systems have arsenic concentrations near the regulatory limit of 10 μg/L, with over two million people potentially exposed to even higher levels from unregulated private wells [264, 265].

Beyond contaminated water, arsenic is present in a variety of foods, including rice (and rice-based products), apple juice, chicken, wine, and beer [266–268]. The US Food and Drug Administration (FDA) has found arsenic in nearly all rice and rice-based baby foods tested [268]. Thus, exposure to inorganic arsenic is widespread and may begin in early childhood [269].

4.15.1 Does Arsenic Cause Obesity?

Given that arsenic and obesity are both linked to diseases such as type 2 diabetes, cardiovascular disease, and certain cancers, researchers have asked whether arsenic might directly contribute to the development of obesity [270, 271].

Animal studies have produced mixed results. Some suggest that arsenic exposure, especially when combined with a high-fat diet, increases body weight and fat deposition [272–274]. Others report the opposite or no effect, with some finding that arsenic exposure reduces fat accumulation or suppresses fat cell development [275, 276]. These inconsistencies may reflect differences in species, dose, timing, diet, or fundamental differences in arsenic metabolism between animals and humans [277, 278].

Human studies have not found consistent evidence that arsenic exposure increases body weight or BMI. Several cross-sectional studies have shown either no association or even inverse associations (higher arsenic linked to lower BMI) [279–287]. A large study in Northern Chile, where historical arsenic exposures reached very high levels, found no relationship between arsenic exposure and being overweight or obese, even after adjusting for age, sex, diet, and other factors [287]. Many studies did not differentiate between toxic inorganic and less toxic organic arsenic forms, and most were cross-sectional, limiting their ability to establish causality.

Hence, there is *little evidence that arsenic directly causes obesity in humans*. Instead, the main concern is that both arsenic and obesity independently increase the risk of similar diseases, and may interact to worsen these outcomes [270].

> **Key Point:** Human evidence does not support a direct role for arsenic in causing obesity. Both exposures are widespread and contribute to overlapping health risks, but arsenic does not explain why people gain weight [270].

4.15.2 Obesity Increases Susceptibility to Diseases

Although arsenic may not cause obesity, **obesity appears to increase susceptibility to the harmful effects of arsenic**, especially for diseases like type 2 diabetes, cardiovascular disease, and certain cancers.

Type 2 Diabetes: Arsenic is a well-established cause of type 2 diabetes, with multiple studies and meta-analyses showing increased risk in populations exposed to elevated arsenic [271, 288–291]. The risk is even higher for obese individuals. For example, in Northern Chile, the increase in diabetes risk associated with high arsenic exposure was over 200% in obese people, compared to about 50% in the general population [288]. Similar findings have been reported in Bangladesh and Taiwan [292, 293].

Cancer and Respiratory Disease: Obesity also amplifies the risk of arsenic-related cancers (lung, bladder) and respiratory symptoms. In Northern Chile, the odds of developing arsenic-related lung or bladder cancer were substantially higher in those with high BMI, with evidence of synergistic (more-than-additive) effects [287]. Obese individuals exposed to arsenic also had much higher risks of respiratory symptoms like cough and shortness of breath [294].

Cardiovascular Disease: Arsenic exposure increases risk for heart disease and hypertension, with some evidence that these risks are greater in those with higher BMI [295, 296].

Experimental Support: Animal studies confirm that arsenic exposure combined with a high-fat diet (a model for obesity) produces worse outcomes than either exposure alone, including greater insulin resistance, liver damage, and kidney injury [297–300].

> **Key Point:** Obesity markedly increases the risk of arsenic-related diseases, including diabetes, cancer, and cardiovascular disease. Arsenic and obesity interact to produce greater harm than either factor alone [270, 287, 288, 294].

4.15.3 What (Not) To Do: Recommendations

○ **Do not assume arsenic causes obesity in humans.** The evidence does not support a direct causal link [270].

○ **Do recognize that arsenic exposure is common, even in developed countries, via water and food.**

○ **Do understand that obesity increases vulnerability to arsenic-related diseases, especially diabetes and cancer.**

○ **Do consider both arsenic exposure and BMI when assessing health risks in populations.**

○ **Do advocate for policies that address both obesity and environmental contaminants.** Current arsenic standards do not account for increased risk in obese populations.

○ **Do support research into how obesity modifies risks from other environmental toxins.**

> **Main Takeaway:** Arsenic does not appear to cause obesity in humans, but obesity greatly increases susceptibility to arsenic-related health risks. Public health efforts should address both exposures together to reduce disease burden [270, 287, 288, 294].

4.16 Coffee or Tea?

Coffee and tea are the world's most popular beverages after water, each with a rich array of bioactive compounds and a large body of research examining their effects on health. While both are often lauded for their potential to reduce inflammation and lower the risk of chronic diseases, their effects are not identical. This section reviews the general health impacts of coffee versus tea, highlights key differences, and offers evidence-based guidance on enjoying these beverages as part of a healthy lifestyle.

4.16.1   Shared Benefits

Large-scale epidemiological studies and meta-analyses consistently show that both coffee and tea consumption are associated with reduced risk of all-cause mortality and cardiovascular death. For tea, a meta-analysis of 18 prospective cohort studies found that higher tea intake was linked to lower mortality from all causes and from cardiovascular disease specifically [301]. Similarly, for coffee, umbrella reviews and large cohort studies demonstrate an inverse relationship between regular coffee consumption and risk of death from any cause and from cardiovascular disease [302].

Both beverages are also associated with a reduced risk of type 2 diabetes. A meta-analysis by Carlström and Larsson found that each additional daily cup of coffee was associated with a 7% reduction in risk of developing type 2 diabetes [303]. For tea, Yang et al. found a similar risk reduction for type 2 diabetes, especially at higher intakes (up to three cups per day) [304]. These effects are likely mediated by improvements in insulin sensitivity, modulation of glucose metabolism, and, to some extent, anti-inflammatory actions.

> **Key Point:** Both coffee and tea are consistently associated with lower risks of cardiovascular disease, type 2 diabetes, and all-cause mortality in large population studies [301–304].

4.16.2   Differences: Mechanisms and Strengths

Both beverages, then, appear to support cardiovascular and metabolic health, but they do so via different mechanisms and with distinct strengths. Coffee consumption seems to have a particularly strong protective effect against liver diseases, including liver fibrosis and metabolic-associated fatty liver disease, with meta-analyses showing a 30–35% reduction in risk for high coffee consumers [305,306]. Coffee is also robustly linked to increases in adiponectin, an anti-inflammatory adipokine, as well as improved cholesterol profiles and marked antioxidant effects [307, 308]. The main active compounds in coffee (namely, caffeine, chlorogenic acid, and diterpenes) exert anti-inflammatory, antioxidant, and metabolic benefits [307].

Tea, particularly green tea, is more strongly associated with improvements in blood pressure, cholesterol, and body weight, and may offer more pronounced antihypertensive and lipid-lowering effects than coffee [309, 310]. Tea's polyphenols are powerful antioxidants that reduce oxidative stress and modulate a broad range of inflammatory pathways [307]. While green tea is generally regarded as more anti-inflammatory than black tea in laboratory and animal studies, large human cohorts have found benefits for both types [311, 312].

> **Key Point:** Coffee may offer greater benefits for liver health and adiponectin, while tea is especially effective for blood pressure, cholesterol, and body weight. Both confer cardiovascular and metabolic protection, but their strengths and mechanisms differ [302, 307, 310].

4.16.3 Inflammation and Biomarkers

The anti-inflammatory effects of both beverages are well documented but modest and sometimes inconsistent in clinical trials. Large cohort and intervention studies show that regular coffee and tea consumption is associated with lower levels of C-reactive protein (CRP, a marker of inflammation and increased cardiovascular/metabolic risk) and higher adiponectin (a marker of anti-inflammatory and metabolic health), particularly in women [308, 311]. Green tea may also reduce LDL cholesterol, and both beverages are associated with decreased oxidative stress.

However, the impact on inflammatory biomarkers such as IL-6 and TNF-α varies by study, population, and dose, and is sometimes absent in short-term interventions. The clinical significance of these changes, compared to other dietary or pharmacological interventions, is modest.

> **Key Point:** Coffee and tea may modestly reduce inflammation and improve markers like CRP and adiponectin, but effects are variable and generally smaller than those seen with dedicated anti-inflammatory drugs or lifestyle changes [307, 311].

4.16.4 Practical Considerations and Confounders

A number of factors influence the health effects of both beverages. The addition of sugar or cream can negate their benefits and may increase inflammatory markers [313]. Preparation methods (such as whether coffee is filtered or unfiltered, or whether tea is green or black) affect the content and bioavailability of active compounds. Moderate consumption, typically two to four cups per day, appears optimal. Effects may also differ by sex, ethnicity, baseline health status, and genetics, particularly in relation to caffeine metabolism. Finally, benefits are most apparent when coffee and tea are consumed as part of a minimally processed, plant-rich diet.

Both coffee and tea can be enjoyed daily as part of a healthy lifestyle. Neither should be relied upon as a sole strategy for reducing inflammation or disease risk, as their effects are modest relative to other lifestyle factors. Avoiding excessive sugar and cream is important to preserve their health-promoting properties [313]. Ultimately, the choice between coffee and tea can be guided by personal preference and individual tolerance, as both have substantial evidence supporting their inclusion in a healthy diet. For those with arrhythmias, anxiety, or who are pregnant, caffeine intake should be moderated. The greatest benefits come from an overall dietary pattern that emphasizes plant-based, whole foods.

Hence, both coffee and tea are health-promoting beverages that can be safely enjoyed in moderation. Coffee is particularly protective for liver health and diabetes, while tea excels in blood pressure and cholesterol control. Either choice can be valuable, and the benefits are most pronounced when these drinks are part of an overall balanced diet [302, 307].

4.16.5 What (Not) To Do: Recommendations

○ **Do enjoy coffee or tea daily as part of a healthy diet.** Both confer significant health benefits for cardiovascular and metabolic health.

○ **Do not rely on either beverage as a sole strategy for reducing inflammation or disease risk.** Their effects are modest compared to other lifestyle factors.

○ **Do avoid excessive sugar and cream.** These additions can offset health benefits and increase inflammation [313].

○ **Do select the beverage you enjoy and tolerate.** Both are beneficial; personal preference and tolerance (e.g., for caffeine) should guide your choice.

o **Be mindful of medical conditions.** Individuals with arrhythmias, anxiety, or pregnancy should moderate caffeine intake.

o **Do focus on overall dietary and lifestyle patterns.** The health benefits of coffee and tea are amplified when combined with a plant-rich, minimally processed diet.

> **Main Takeaway:** Both coffee and tea are health-promoting beverages that can be safely enjoyed daily in moderation. While their benefits differ slightly (e.g., coffee is especially protective for liver health and diabetes, tea excels in blood pressure and cholesterol control) either can be a valuable part of a healthy lifestyle. Choose the drink you prefer and enjoy the benefits, knowing that both contribute to long-term health when consumed as part of an overall balanced diet [302, 307].

4.17 How to Lose Stomach Fat: Alcohol

When it comes to losing stomach fat, it's vital to understand that not all body fat is the same. We have subcutaneous fat, which sits just beneath the skin, and visceral fat, which is stored deep inside the abdomen, surrounding vital organs like the liver, pancreas, and intestines. Visceral fat is particularly harmful, as it is strongly linked to increased risks of type 2 diabetes, heart disease, and certain cancers. Reducing this type of fat is therefore not just a matter of aesthetics, but a critical step for long-term health.

4.17.1 Alcohol and Visceral Fat

A growing body of research demonstrates that alcohol consumption is closely linked to increased levels of visceral fat; even in people who are otherwise normal weight. This association has been observed across diverse populations and is independent of overall body mass index (BMI) [314–318].

Large cross-sectional studies in Korea and Europe have shown that the amount of alcohol consumed per drinking occasion is a strong predictor of abdominal obesity, even among individuals with a normal BMI. For example, in a nationally representative Korean sample of over 11,000 normal-weight adults, those who consumed seven or more drinks per occasion had significantly higher odds of abdominal obesity compared to those who drank less, regardless of how often they drank [314]. Binge drinking (defined as consuming seven or more drinks for men, five or more for women, on one occasion) was also associated with higher risk of abdominal obesity, especially in men who binged daily. Notably, the frequency of drinking alone was not associated with abdominal obesity; rather, it was the quantity consumed per occasion that mattered most [314].

Similar findings have been reported in Western populations. In a study of elderly Swedish men, higher alcohol intake was associated with greater waist circumference and waist-to-hip ratio, but not with overall BMI [315]. This suggests that alcohol promotes the accumulation of visceral (abdominal) fat rather than generalized weight gain. Other studies have confirmed that heavy or binge drinking is linked to increased central adiposity, even after adjusting for age, physical activity, smoking, and total energy intake [316–318].

Testosterone plays a vital role in regulating body composition in men. It promotes muscle growth and supports the breakdown of fat, especially visceral fat. Chronic alcohol consumption has been shown to lower testosterone levels by directly damaging the cells in the testes responsible for its production [319, 320]. Lower testosterone not only impairs muscle building but also makes it harder to lose fat, particularly from the abdominal region.

> **Key Point:** Alcohol intake, especially in large quantities per occasion or through binge drinking, is strongly associated with increased visceral (abdominal) fat; even in people who are not overweight by BMI.

4.17.2   Why Does Alcohol Increase Visceral Fat?

There are several mechanisms by which alcohol may promote the accumulation of visceral fat:

o **Caloric Density:** Alcohol contains 7 calories per gram, making it nearly as energy-dense as fat, but with no nutritional value. Many alcoholic beverages are also high in added sugars, further increasing calorie intake [315, 316].

o **Metabolic Effects:** Alcohol is metabolized primarily in the liver, where it can disrupt normal fat metabolism. Chronic alcohol intake impairs the liver's ability to oxidize fat, leading to increased fat storage, particularly in the abdominal region [315, 318].

○ **Hormonal Disruption in Men:** Alcohol damages cells in the testes, reducing testosterone production [319, 320]. Testosterone is a key hormone for muscle formation and fat breakdown; lower levels are associated with increased visceral fat and reduced muscle mass.

○ **Endocrine Changes:** Alcohol consumption can increase cortisol (a stress hormone), which is known to promote fat accumulation in the abdomen [315, 318].

○ **Behavioral Factors:** Alcohol lowers inhibitions and may lead to increased food intake, particularly of high-calorie, unhealthy foods.

4.17.3 How Much Alcohol Is Too Much?

The evidence suggests that even modest habitual alcohol consumption (more than 5 units (drinks) per week) can have adverse effects on semen quality and reproductive hormones, with more pronounced effects at higher intakes [320]. For abdominal fat, the risk increases sharply with higher quantities per drinking occasion and with frequent binge drinking [314–316].

4.17.4 What (Not) To Do: Recommendations

○ **Do not ignore alcohol as a contributor to belly fat.** Even if your weight is normal, alcohol can increase visceral fat and associated health risks.

○ **Do limit the quantity of alcohol per occasion.** The amount you drink at one time is more important than how often you drink.

○ **Do avoid binge drinking.** Binge drinking is strongly linked to increased abdominal fat, especially in men.

○ **Do not rely on BMI alone to assess your health risk.** Waist circumference and other measures of abdominal fat are better indicators of metabolic risk.

○ **Do consider cutting back or abstaining from alcohol if you want to lose belly fat.** Reducing alcohol intake is one of the most effective strategies for reducing visceral fat.

○ **Be mindful of hidden calories in alcoholic beverages.** Many drinks are high in sugar and calories, which can undermine your fat loss efforts.

Main Takeaway: If you want to lose belly fat, cutting down on alcohol is one of the most evidence-based steps you can take. Alcohol promotes the accumulation of harmful visceral fat through multiple mechanisms, including increased calorie intake, disrupted fat metabolism, and lowered testosterone. Even moderate drinking can have an impact, especially if you tend to drink larger quantities per occasion. For optimal health and fat loss, limit alcohol intake and focus on sustainable, healthy lifestyle changes.

Chapter 5

The Battle for Truth in Nutrition Science

Navigating the truth in nutrition and dietetics often means balancing genuine scientific evidence with the powerful pull of popular beliefs. In the realm of dietary advice, this tension takes on heightened significance: the difference between evidence-based recommendations and persuasive anecdotes can directly impact health outcomes. History provides ample cautionary tales (from misguided medieval dietary cures to the widespread promotion of miracle foods and fad diets) that illustrate our tendency to embrace narratives aligning with our preconceptions, even when research tells a different story. In the digital age, social media and online communities intensify this effect, fueling confirmation bias and turning nutrition debates into arenas where open-minded inquiry is frequently at odds with entrenched dogma.

Genuine pursuit of nutritional truth requires embracing complexity rather than oversimplification. Just as medieval societies wrongly explained plague deaths through superstition rather than understanding the true biological causes, people today often fall into the trap of reducing nuanced dietary science to oversimplified debates; such as "clean eating" versus "processed foods," or "superfoods" against "dangerous additives." These black-and-white thinking patterns persist in part because of enduring myths, like the long-held belief in spinach's extraordinary iron content, which originated from a simple typographical error but influenced nutrition advice for decades. Such stories endure not due to scientific validity, but because they comfortably fit into established beliefs and cultural narratives.

Distinguishing nutritional fact from fiction becomes especially challenging when genuine evidence collides with conspiracy-laden viewpoints. In cases where people interpret the same nutrition studies differently (debating, for example, the health impacts of carbohydrates or fats) meaningful discussions can occur, and consensus may be reached through careful review of the data. However, dialogue breaks down when one group disregards established dietary science in favor of fabricated claims; such as insisting that mainstream nutrition advice is part of a grand cover-up, or promoting miracle diets based on entirely invented evidence. In these situations, the opportunity for finding common understanding disappears, since constructive conversation relies on a shared commitment to reality, not on engaging with misleading or wholly fictional information.

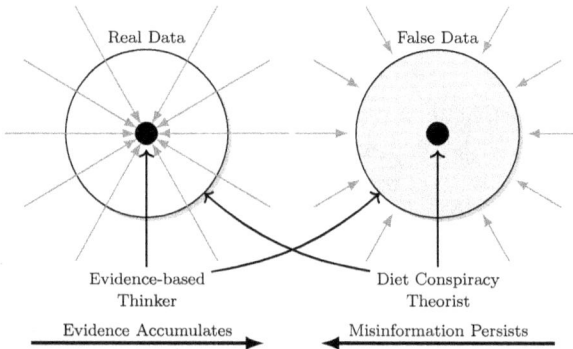

Figure 5.1: This graphic depicts the stark contrast between how evidence-based thinkers and diet-related conspiracy theorists interpret nutritional information. On the left, the "Evidence-Based Thinker" gathers findings from scientific studies, expert consensus, and peer-reviewed research, continuously refining their understanding of nutrition as new evidence emerges. On the right, the "Diet Conspiracy Theorist" relies on misinformation and disregards scientific consensus, favoring sensational or unsubstantiated dietary claims that reinforce existing biases. This illustration highlights the difficulty of bridging these perspectives, as conspiracy-driven views often reject the core principles of scientific inquiry, making meaningful discussion about nutrition nearly unattainable.

Consider, for instance, the enduring debates around dietary guidelines and nutrition science. Just as vaccines have revolutionized public health, evidence-based dietary recommendations (such as the benefits of balanced macronutrient intake or the dangers of trans fats) have been repeatedly validated through rigorous research. Yet, similar to early resistance to the germ theory of disease or the initial dismissal of asbestos risks, some individuals continue to reject well-established nutritional science simply because its effects are not always immediately visible or intuitively obvious to them. For example, the risks associated with excessive sugar consumption or ultra-processed foods were once underestimated, only becoming clear through years of accumulating scientific evidence; not through speculation or misinformation. As shown in Figure 5.1, the way people filter and accept dietary information often depends on confirmation bias: many are inclined to accept only those nutritional claims that fit their preexisting beliefs, while disregarding robust scientific data that challenges their views.

The gray arrows in Figure 5.1 demonstrate how people can interpret the same nutritional research in different ways, leading to a range of opinions and dietary choices. This diversity of interpretation is a natural aspect of how we process information, shaped by personal biases, cultural influences, and previous experiences with food and health. Importantly, these varying perspectives are all anchored in the same set of scientific findings, which enables healthy debate and the possibility of reaching consensus through ongoing research and open discussion.

The real difficulty arises when nutritional beliefs are built on misinformation or fabricated data. In these situations, genuine conversation stalls because there is no shared factual basis to reference. For any meaningful dialogue about diet and health to take place, all participants must rely on the same credible, observable evidence. Without this common ground, discussion becomes futile, and the only way to move forward is for everyone involved to return to evidence-based information and engage with the realities of nutrition science.

The challenge in bridging the divide between evidence-based nutrition and conspiracy-driven dietary beliefs stems from a fundamental rejection of scientific evidence by the latter. While history has shown that certain controversial health concerns (such as the dangers of trans fats) were once dismissed before being substantiated by solid research, their eventual acceptance was always grounded in robust, observable data, not in speculation or fabricated claims. This vital difference makes it extremely difficult to foster productive dialogue between those who trust in scientifically validated nutrition advice and those who promote dietary conspiracies unsupported by empirical evidence.

In the world of nutrition, certain influencers present their skepticism of mainstream dietary guidelines as a form of "resistance" against powerful or shadowy interests, often tapping into past instances of institutional betrayal.

Their followers, convinced they are more discerning than those who accept established nutrition science, frequently echo influencer slogans like "wake up" or "think for yourself," all while dismissing rigorous, peer-reviewed nutritional research as biased or corrupt. Ironically, this groupthink mimics the very uncritical acceptance they claim to reject.

This dynamic is reminiscent of the misinformation spread during the trans fat controversy or misleading marketing behind fad diets, where commercial interests at times overshadowed scientific consensus. As with the aftermath of the asbestos scandal, communities with a history of being misled by institutions may become especially vulnerable, equating skepticism with true critical thinking and mistaking mere contrarianism for genuine insight.

Translating nutrition science into accessible language is vital, yet it carries the risk of oversimplifying complex realities. This balancing act is particularly critical in the field of diet and health, where misunderstandings can quickly translate into harmful choices. With so much at stake, the proliferation of nutrition myths and misinformation can contribute to risky dietary behaviors and adverse health outcomes, as the following case studies demonstrate.

> **Key Point:** Productive dialogue about diet and health is only possible when all participants rely on credible, evidence-based information. When dietary beliefs are built on misinformation or fabricated claims, meaningful discussion breaks down; highlighting the necessity of a shared factual foundation for informed choices and public trust in nutrition science.

5.1   The Vital Role of Science Popularization

Bringing nutrition science to the public is vital for empowering individuals to make healthier choices. Translating the intricate language of nutritional research into clear, relatable information serves as an essential bridge between scientists and everyday people. This process not only sparks curiosity about diet and health but also encourages scientific literacy, critical thinking, and more informed decision-making about what we eat and how we live.

However, communicating nutrition science to a broad audience isn't without its pitfalls. A major challenge lies in striking the right balance between making information accessible and preserving its accuracy. When nutrition facts are overly simplified, they can easily become distorted, leading to widespread misunderstandings; such as the demonization of all fats or the elevation of single "superfoods" as cure-alls.

Additionally, the desire to capture attention can sometimes result in exaggerated or sensationalized headlines about diets and healthy eating. This can foster confusion, erode public trust in nutrition science, and even open the door to conspiracy theories about dietary guidelines or food safety. In the realm of diet and health, such misinformation doesn't just cause confusion; it can directly influence behaviors that result in poor health outcomes or serious harm.

> **Key Point:** Effective nutrition communication requires presenting evidence-based information in a way that is both accessible and accurate. Oversimplification or sensationalism can distort public understanding and lead to harmful dietary choices, underscoring the need for clear, honest, and nuanced messaging that empowers people to make informed decisions about their health.

5.1.1 The Subtle Deception of (Mis)information

Communicating nutrition science to the public is vital for supporting informed dietary choices, but the line between simplifying and distorting information is razor-thin; especially in the digital age. Even small shifts in how nutrition advice is presented can quickly spiral into widespread misconceptions, eroding trust in evidence-based dietary guidance.

A notable example is a 2021 study that surveyed individuals who had followed a meat-only diet for at least six months, asking them to report on their experiences [321]. This methodology inherently excludes anyone who tried the diet but discontinued it earlier due to negative effects, thus biasing the results toward positive outcomes. This is not just a theoretical concern, there are specific, documented reports of people experiencing adverse symptoms such as dizziness, brain fog, hair loss, bleeding gums, and other unpleasant effects well before reaching the six-month threshold. Those individuals had to stop the diet before the six months due to these negative experiences and, as a result, were systematically excluded from the study's sample. By setting the inclusion criteria at six months, the study omits these negative experiences, creating a misleadingly favorable impression of the diet.

Online, influencers who promote such restrictive diets often conveniently omit these critical details. They don't mention the study's exclusion of early dropouts or the abundance of negative short-term reports, instead presenting the findings as robust evidence of the diet's safety and effectiveness. Frequently, these influencers base their claims on debunked or poorly designed studies, or, in cases where no credible evidence of harm exists (such as with seed oils, where research consistently shows neutral or beneficial effects) they simply ignore the science and assert the opposite.

The reason for this is simple: contrarian content drives engagement. Sensational claims and viral videos attract more clicks, shares, and followers than careful, evidence-based explanations. More clicks translate directly into more revenue, incentivizing influencers to prioritize controversy over accuracy. As a result, extreme and misleading narratives about diets and foods gain traction, while nuanced, science-backed advice is less followed.

This cycle of oversimplification and sensationalism not only spreads misinformation but also deepens public distrust in nutrition science. It highlights the urgent need for critical digital literacy and skepticism when evaluating dietary claims online. Without these skills, even well-intentioned individuals can fall prey to cleverly manipulated narratives—potentially making choices that harm their health and further erode trust in legitimate nutrition research.

> **Key Point:** In nutrition science, the way information is framed and shared can profoundly shape public perception and behavior. Critical digital literacy and an understanding of research limitations are essential for discerning credible dietary advice from manipulated or incomplete narratives; helping to safeguard both personal health and trust in evidence-based nutrition.

5.2   Risks in Nutrition Communication

While innovations like interplanetary rockets capture the public's imagination, they often feel distant from everyday concerns. People don't question rocket science. Nutrition and health, on the other hand, have a direct and immediate impact on people's lives, making the public especially attentive (and sometimes highly critical) of claims in this area.

During the COVID-19 pandemic, for example, the case of ivermectin highlighted the dangers of miscommunication in health science. Early lab studies indicated ivermectin could inhibit the SARS-CoV-2 virus in cell cultures [322], but these results did not translate to safe or effective human treatment; the necessary dosages would be toxic [323]. Despite this, headline-grabbing social media posts and news articles frequently stripped away the crucial context, fueling widespread belief that ivermectin was a miracle cure. This led many to self-medicate, often with veterinary products, resulting in preventable health emergencies and even fatalities [324].

Similar dynamics are at play in nutrition communication. Oversimplified or sensationalized claims (such as labeling foods as unequivocally "superfoods" or "toxins") can lead people to adopt unsafe diets, misuse supplements, or avoid beneficial foods based on distorted interpretations of preliminary research. When context and nuance are sacrificed for catchy headlines, misinformation about nutrition spreads rapidly, sometimes with harmful consequences for public health.

This chapter explores how oversimplified nutrition messaging can distort understanding, corrode trust in science, and have tangible negative effects. In a world overflowing with information, responsible and nuanced communication is more important than ever. Within the realm of nutrition and diet, where choices directly affect well-being, the stakes are particularly high: misrepresenting scientific findings can endanger lives, fuel conspiracy thinking, and undermine confidence in health guidance and institutions.

> **Key Point:** Unlike distant fields of science, nutrition and health communication directly shapes real-life choices and outcomes. Oversimplified or sensationalized messages can easily lead to harmful behaviors and erode trust in evidence-based guidance; underscoring the critical need for clear, accurate, and responsible nutrition communication.

5.3 Historical Parallels

This pattern of misconnecting causality extends far beyond dietary myths. In medieval Europe, during the Black Death, communities tragically misunderstood correlations they observed [325]. For example, households with cats (often cared for by women later accused of witchcraft) saw fewer plague cases because cats kept the rat population (and thus plague-carrying fleas) in check [326]. Instead of recognizing this protective effect, fear and superstition led to the scapegoating of these women as witches, resulting in widespread persecution and tragedy [327].

This same cognitive error (i.e., confusing correlation with causation) persists in modern nutrition discourse. Today, many dietary myths and conspiracies arise from a lack of understanding about how to properly interpret scientific evidence. For instance, people may notice that individuals following a certain diet appear healthier, without accounting for confounding factors or selection bias, and then attribute causality where none exists. Likewise, sensational stories about foods being "miracle cures" or "toxic poisons" often gain traction not because of robust evidence, but because they provide a simple narrative that resonates with existing fears or beliefs.

Just as medieval societies blamed witches to make sense of the plague, modern communities can fall prey to nutrition conspiracies when scientific literacy is lacking and communication is unclear. In both cases, societal anxieties and gaps in understanding fuel a search for scapegoats; whether it's "dangerous" foods, industry "villains," or health authorities. When uncertainty in nutrition science is not addressed with transparency and empathy, it leaves room for misinformation to fill the void, perpetuating cycles of blame rather than fostering rational, evidence-based dialogue.

> **Key Point:** Confusing correlation with causation (whether in history or in modern nutrition) can lead to harmful myths, scapegoating, and misguided choices. Promoting scientific literacy and transparent communication is essential to break this cycle, allowing people to make informed decisions based on evidence rather than fear or oversimplified narratives.

5.4   Shifting Nutrition Science

A common criticism directed at nutrition science is that it has been "wrong" many times before, with skeptics pointing to past mistakes as evidence that all dietary recommendations are inherently unreliable. One of the most cited examples is the shifting advice about eggs: for years, the public was told to avoid eggs due to concerns about cholesterol and heart health, only for later research to suggest that, for most people, moderate egg consumption does not significantly increase cardiovascular risk [328]. This back-and-forth has left many people feeling confused and frustrated, fueling the perception that nutrition science is fickle or untrustworthy.

However, this argument fails to distinguish between two very different types of error: deliberate deception for profit and the natural evolution of scientific understanding. The first (e.g., corporate or institutional betrayal) occurs when companies or authorities intentionally hide or distort evidence to protect financial interests. The second (i.e., empirical refinement) is actually science's greatest strength, reflecting its willingness to update recommendations as new evidence emerges.

Cases of institutional betrayal are stark reminders of the consequences when profit trumps public health. Take, for example, the asbestos debacle: despite early signs by the 1930s that factory workers were suffering from lung diseases, companies suppressed the evidence for decades to keep selling their "miracle" material. Similarly, the marketing of OxyContin as non-addictive, despite evidence to the contrary, was driven by corporate interests, not scientific uncertainty. These episodes are not indictments of the scientific method; rather, they expose the dangers of allowing commercial interests to override transparency and ethics. In the nutrition realm, the sugar industry famously funded research in the 1960s and 70s to downplay sugar's health risks while shifting the blame to dietary fat [329]; another case where public trust in nutrition science was undermined by profit-driven manipulation, not honest error.

In contrast, science's process of self-correction is what makes it reliable in the long run. The evolving advice about eggs is a prime example of this process in action. Early studies linked dietary cholesterol from eggs to increased heart disease risk, leading to widespread recommendations to limit egg consumption [328]. However, as research methods improved and larger, more rigorous studies were conducted, scientists found that for most people, dietary cholesterol has a much smaller impact on blood cholesterol than previously thought. This led to updated guidelines that reflect the best available evidence, even if it means reversing earlier advice. While this can be confusing, it is a hallmark of scientific rigor, not a sign of failure.

Consider also the enduring myth that spinach is an exceptional source of iron; a misconception that began with a simple decimal error, vastly overstating its iron content. Although this mistake lingered in popular culture, scientific research steadily corrected the record, clarifying that spinach's iron is both less abundant and less bioavailable than previously believed. This evolution reflects science's capacity for empirical refinement: old assumptions are replaced as new evidence accumulates. The same process is seen in revisions to dietary guidelines as understanding of fats, cholesterol, and carbohydrates has advanced; these updates are not failures, but hallmarks of scientific rigor.

Sometimes, scientific fallibility is wrongly conflated with outright fraud. The infamous retraction of Andrew Wakefield's 1998 study linking vaccines to autism is often cited as proof that science "can't be trusted." In reality, the fraudulent nature of the research was exposed by further studies within a few years, but institutional reluctance to retract the paper allowed the misinformation to persist. The delay was a failure of communication and oversight, not of the scientific process itself. In nutrition, similar confusion emerges when poorly designed or manipulated studies are used by influencers to promote fad diets, while robust evidence is ignored or dismissed.

It's vital to separate intentional deception from the natural, necessary evolution of nutrition science. When dangerous products like Radithor's radioactive tonic caused harm, it was not due to a lack of scientific understanding about radiation's risks, but the absence of proper regulation and ethical oversight. Conversely, the transparent revisions of dietary guidelines over time (or the rapid sharing of new findings, as seen with mRNA vaccine research) demonstrate science's strength: adaptability in the face of new knowledge. Ultimately, labeling all scientific shifts as "failures" overlooks the reality that progress depends on questioning, correction, and transparency. Science's so-called "wrongness" is only damaging when driven by greed or obfuscation; when rooted in honest inquiry, it is the engine of genuine progress in nutrition; and in all fields that touch public health.

> **Key Point:** Scientific progress in nutrition depends on continual questioning, correction, and transparency. While evolving recommendations may seem confusing, they reflect the self-correcting nature of science; not its failure. Distinguishing between deliberate deception and honest refinement is vital for maintaining trust in nutrition science and making informed dietary choices [328, 329].

5.5 Evidence-Based Nutrition Science

While previous sections have examined the origins of skepticism toward nutrition science (from historic scapegoating to modern scandals driven by profit) it's vital to acknowledge the extraordinary, quantifiable benefits that evidence-based approaches have brought to public health and nutrition. Few achievements illustrate this more clearly than the dramatic reduction in child mortality worldwide over the last two centuries.

As depicted in Figure 5.2, the global mortality rate for children under five has fallen from around 45% in 1800 to just 3.7% in 2020. This remarkable progress is not due to a single breakthrough, but rather to the steady accumulation of scientific advances in both medicine and nutrition. Key milestones include the introduction of vaccinations (such as Jenner's smallpox vaccine in 1796), the acceptance of germ theory (thanks to pioneers like Pasteur and Koch), improvements in sanitation and access to clean water, the discovery and use of antibiotics, and widespread adoption of oral rehydration therapy.

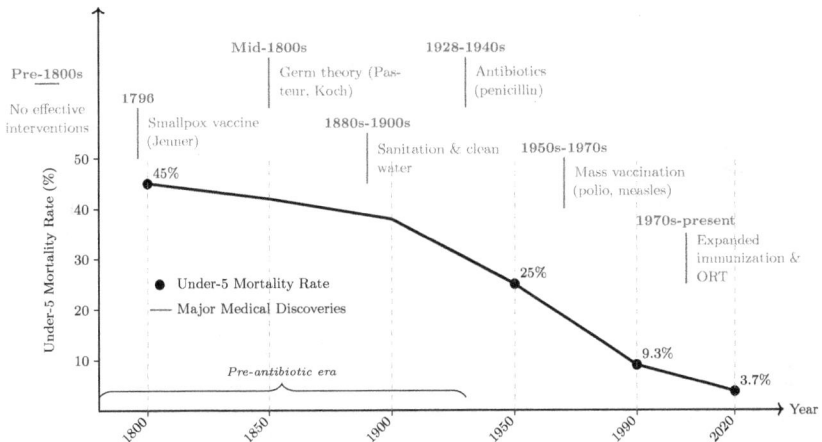

Figure 5.2: The dramatic decline of child mortality through medical progress, 1800-2020. This figure illustrates the remarkable reduction in under-5 mortality rates from 45% in 1800 to 3.7% in 2020, coinciding with major medical breakthroughs including vaccination (1796), germ theory (mid-1800s), sanitation improvements (1880s-1900s), antibiotics (1928-1940s), mass vaccination programs (1950s-1970s), and expanded immunization with oral rehydration therapy (ORT) (1970s-present). The steepest mortality decline occurred after 1950, marking the transition from the pre-antibiotic era to modern evidence-based medicine.

Nutrition interventions have played a major role in this success story. The development of fortified foods, widespread vitamin supplementation, advances in infant and maternal nutrition, and targeted programs to combat malnutrition have all contributed to the steep decline in child mortality. These achievements underscore the immense value of evidence-based nutrition and public health strategies; demonstrating that, despite understandable skepticism, the scientific process remains our most effective tool for improving lives on a global scale.

A powerful example of this is the fortification of crops with folic acid. Folic acid, a B-vitamin, is essential for healthy fetal development, particularly in preventing neural tube defects (i.e., serious birth defects that affect the brain, spine, or spinal cord of a developing fetus). In the late 1990s and early 2000s, many countries began mandating the addition of folic acid to staple foods like wheat flour and cornmeal. In some regions, folic acid was even sprayed directly onto crops or added during food processing to ensure widespread coverage. This public health intervention was based on robust evidence showing that increased folic acid intake among women of childbearing age dramatically reduced the incidence of neural tube defects in newborns [330].

The results were striking: after the introduction of mandatory folic acid fortification, countries such as the United States, Canada, and Australia saw a significant drop in neural tube defects and related infant mortality. This intervention is a clear demonstration of how evidence-based nutrition policy (grounded in rigorous research and implemented at scale) can lead to measurable improvements in population health. Such examples highlight the transformative power of nutrition science when it is guided by evidence and applied through thoughtful public health strategies.

Ironically, the very success of these interventions has contributed to a new wave of misinformation. Now that neural tube defects have become so rare in many countries, thanks to decades of effective folic acid fortification, the public is less familiar with the devastating consequences of folic acid deficiency. As a result, people are increasingly exposed to fearmongering videos and social media posts that exaggerate or fabricate risks associated with folic acid, often ignoring the overwhelming evidence of its benefits. This phenomenon highlights a recurring challenge in nutrition communication: when a public health intervention is so effective that it makes a disease nearly invisible, it can become an easy target for misinformation and conspiracy theories. The dramatic reduction in neural tube defects is a testament to the power of science-driven policy, and serves as a reminder that the benefits of such interventions can far outweigh the unfounded fears that sometimes circulate online.

Oral rehydration therapy (ORT) stands as one of the most impactful yet often overlooked public health achievements of the late 20th century; an achievement built on simple nutritional science. Introduced in the 1970s, ORT involves administering a carefully balanced mixture of clean water, glucose, and essential electrolytes (primarily sodium and potassium). This straightforward solution can be given at home or in clinics to children suffering from dehydration caused by severe diarrhea; a major cause of child mortality worldwide. Prior to the advent of ORT, acute diarrheal diseases frequently resulted in life-threatening dehydration, especially in regions with limited access to medical care. The brilliance of ORT is in its accessibility and ease of use: it does not require advanced medical equipment or costly facilities, just fundamental ingredients and clear guidance.

ORT's life-saving impact is undeniable. Since its widespread introduction, tens of millions of lives have been saved, with dramatic reductions in deaths from conditions such as cholera, rotavirus, and other diarrheal diseases, particularly in low- and middle-income countries. ORT's success highlights how nutrition-based, evidence-driven interventions can transform global health—not through expensive innovation, but through practical, scalable solutions that empower communities. This example demonstrates the profound potential of nutrition science when it is translated into accessible, evidence-based public health strategies. It is a reminder that some of the most effective answers to complex health challenges are not high-tech or costly, but rooted in a clear understanding of basic nutritional needs and the power of community education.

Many of the most important advances in nutrition and public health were initially met with skepticism, resistance, or even outright hostility; mirroring the patterns of mistrust and misattribution explored earlier in this chapter. Whether it was the introduction of food fortification, the rollout of vaccines, or the adoption of oral rehydration therapy, each breakthrough faced initial pushback from the public, policymakers, or even segments of the scientific community. Yet as robust, real-world evidence accumulated, the positive impact of these interventions became impossible to deny. The most dramatic reductions in child mortality followed the widespread adoption of antibiotics, vaccines, and nutrition-based interventions in the mid-20th century, signaling the shift from anecdotal, tradition-based practices to modern, evidence-based approaches to health and nutrition.

This history highlights a central paradox: cycles of skepticism, exploitation, and misinformation have repeatedly threatened to erode confidence in nutritional and medical science, yet the long-term data tell a clear story of remarkable progress. In fact, many of the very interventions now targeted by conspiracy theories (such as food fortification, vaccines, and improved sanitation) are the same ones responsible for saving hundreds of millions of lives.

The takeaway is clear: while science is not infallible, its transparent, self-correcting nature is precisely what makes it so powerful and transformative. The challenge for our time is to ensure that lingering fear and distrust do not overshadow the real, measurable gains made through evidence-based nutrition and public health initiatives, nor inhibit continued progress in improving global well-being.

> **Key Point:** The most profound improvements in global health (including dramatic reductions in child mortality) have come from evidence-based nutrition and public health interventions. While skepticism and misinformation may challenge progress, the self-correcting nature of science ensures that, over time, rigorous research and transparent policy lead to real, measurable benefits for society. Trust in this process is essential for continued advancement in nutrition and well-being.

5.6   Contrarianism vs. Critical Thinking

In today's digital landscape, evidence-based nutrition and health information often receives far less attention than sensational or contrarian claims. The economics of online platforms reward content that challenges established knowledge, making it far more profitable to promote controversial or surprising viewpoints than to highlight careful, consensus-driven science.

A recurring cognitive distortion in nutrition discourse is the tendency to equate contrarian attitudes with genuine critical thinking. While healthy skepticism is crucial for scientific progress, contrarian narratives frequently replace evidence-based analysis with ideological opposition. This is especially common in communities prone to conspiracy thinking, where skepticism is framed as intellectual independence or superior insight, and mainstream scientific consensus is dismissed as corrupt or misguided. Recent psychological research shows this dynamic is particularly pronounced among individuals with narcissistic personality traits, such as grandiosity and a need for uniqueness, which can interact with education in complex ways to heighten susceptibility to conspiracy beliefs [331].

The link between narcissistic traits and belief in nutrition conspiracies (such as anti-seed oil or anti-fortification movements) often stems from epistemic imbalance: a mismatch between self-perceived knowledge and actual evidence-based understanding. In these situations, people may feel deeply confident in their dietary opinions, even when unsupported by scientific evidence or critical analysis. This overconfidence makes them more vulnerable to misinformation, as the sense of "knowing" outweighs what is truly supported by research [331].

Studies have shown that narcissists frequently display high intellectual self-confidence but relatively weak critical thinking skills, making them especially susceptible to pseudoscientific or conspiratorial dietary claims [332, 333]. Interestingly, higher levels of formal education do not always protect against these beliefs. In a large-scale study of more than 51,000 participants, Cosgrove and Murphy (2023) found that while cognitive reflection (the ability to question intuitive but incorrect responses) reduced conspiratorial thinking, higher education actually strengthened the relationship between narcissism and belief in COVID-19 conspiracies [331]. This suggests that, without explicit training in critical reasoning, education can simply increase confidence in one's possibly flawed beliefs.

This pattern is reminiscent of historical episodes where ideological rigidity replaced systematic investigation. During medieval plagues, communities often blamed illness on witchcraft rather than natural causes. Today, similar dynamics appear when people reject nutrition consensus (such as the benefits of food fortification or the safety of certain additives) while embracing alternative narratives without critical scrutiny. The tendency to treat dissent as insight, while ignoring the evidence base, demonstrates a misapplication of critical thinking.

Cognitive reflection stands out as a protective factor against these distortions. The aforementioned large-scale study found that individuals with higher cognitive reflection scores were less likely to endorse conspiracy beliefs, regardless of their narcissistic tendencies. This analytic skill (i.e., the ability to pause and reconsider initial impressions) helps counteract the overconfidence and selective doubt characteristic of conspiratorial contrarianism. Interventions that foster Bayesian reasoning, intellectual humility, and critical reflection may prove more effective at combating nutrition misinformation than traditional science education alone.

The consequences of confusing contrarianism for critical thinking go beyond individual misunderstanding. Conspiracy-driven narratives offer the illusion of expertise: mastering nutrition science requires time and effort and engagement with complex evidence, while it is much easier to claim that experts are simply deceiving the public. This shortcut can be self-reinforcing, leaving individuals more susceptible to manipulation by those with ulterior motives. Addressing these challenges will require systemic changes that prioritize critical reasoning, transparency, and robust evaluation of both information sources and digital platforms.

> **Key Point:** Contrarianism is not the same as critical thinking. While skepticism is a vital component of scientific inquiry, true critical thinking in nutrition requires evaluating evidence impartially, recognizing one's own biases, and updating beliefs in light of new data. Educational approaches that promote cognitive reflection and intellectual humility may be more effective than rote instruction in reducing susceptibility to nutrition misinformation and conspiracy thinking [331–333].

5.7   Educational Paradox

The connection between formal education and critical thinking in nutrition is more nuanced than commonly believed. While many assume that advanced education automatically translates to better dietary decisions, emerging research suggests the reality is far more complex. In a 2023 study, Cosgrove and Murphy uncovered a surprising trend: among individuals with narcissistic personality traits, higher levels of education actually correlated with a greater tendency to believe in nutrition-related conspiracy theories; a phenomenon they describe as the "educational paradox" [331]. In essence, academic credentials can foster a sense of expertise without necessarily enhancing the ability to critically evaluate evidence, enabling some to use technical language to dismiss well-supported nutritional science.

This paradox is particularly prevalent within nutrition and dieting circles. It's not uncommon for individuals with advanced degrees in fields unrelated to nutrition to use their credentials to bolster dietary claims that diverge from the scientific consensus. For example, self-proclaimed "nutrition experts" with doctorates outside of nutrition or dietetics (such as chiropractors, for instance) frequently promote extreme regimens like all-meat (carnivore) diets or detox protocols. They often use scientific-sounding language to cloak their recommendations as "skeptical inquiry," while sidelining the robust clinical and epidemiological evidence supporting balanced diets. These practitioners routinely leverage their "doctor" or professional titles to market dietary advice and products that lack scientific backing. Casting doubt on mainstream nutrition advice becomes a strategic move, especially when selling branded supplements or exclusive diet plans.

Followers of these influencers (often themselves educated professionals with high narcissistic tendencies) may enthusiastically adopt slogans such as "do your own research," yet routinely disregard peer-reviewed studies and established nutritional guidelines (see Figure 5.3). This cycle strongly resembles historical patterns, like those seen in the early 20th century when patent medicine promoters used anecdotal stories and claims of authority to market unregulated weight loss tonics, all while avoiding empirical validation.

By understanding this "educational paradox," we can better recognize why formal education alone is not a safeguard against nutrition misinformation; especially when confidence, status, and commercial interests are at play [331]. There are several mechanisms by which the educational paradox operates in the context of nutrition and dieting:

1. **Credentialed Arrogance**: Individuals with advanced education outside of nutrition may conflate expertise in their own field with universal competence. For example, a chiropractor promoting a fad diet may misuse statistical concepts (like "correlation versus causation") to challenge robust findings in nutrition science, despite lacking relevant training.

2. **Selective Literacy**: Education systems that emphasize memorization over critical appraisal can leave gaps in the ability to evaluate scientific claims. A person may master technical skills (such as data analysis) but remain ill-equipped to assess the validity of nutritional research, making them vulnerable to persuasive but inaccurate dietary rhetoric.

3. **Tribal Identity**: Academic degrees become status symbols within online diet communities. Influencers with advanced titles often present unconventional or fringe dietary advice as "suppressed truths," exploiting the deference that lay audiences give to academic authority.

It is often apparent to nutrition scientists and clinicians when individuals with unrelated credentials are misusing their authority to promote misinformation. Years of discipline-specific training are replaced by confidence derived from status in another field, often without self-awareness of the harm caused. This dynamic not only undermines public trust in evidence-based nutrition but illustrates how knowledge without epistemic humility can fuel dietary misinformation. Just as patent medicine salesmen once used titles and testimonials to market ineffective weight loss tonics,

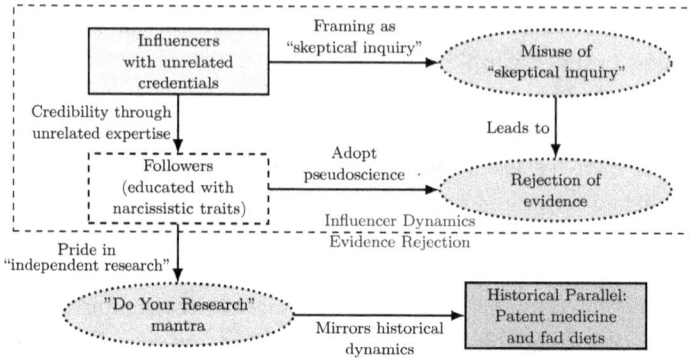

Figure 5.3: The "educational paradox" in nutrition communities, as identified by Cosgrove and Murphy (2023) [331]: influencers with unrelated credentials use academic authority to promote pseudoscientific dietary claims, while their followers adopt contrarian rhetoric and reject peer-reviewed evidence. This cycle, seen in vaccine debates, fad diets, and supplement marketing, mirrors historical patterns of authority misuse in health and nutrition.

today's credentialed influencers exploit academic authority to legitimize dietary myths. Breaking this cycle requires educational systems and public discourse that value curiosity, critical appraisal, and humility over mere credentialism.

> **Key Point:** The "educational paradox" reveals that higher education does not guarantee critical thinking in nutrition. Without epistemic humility and training in evidence appraisal, academic credentials can be misused to spread dietary misinformation. True expertise in nutrition requires both rigorous scientific training and the willingness to question one's own assumptions [331].

As a point of transparency, it should be noted that this very book was authored by someone who is not a credentialed nutritionist. However, unlike the aforementioned cases, this book is not intended to serve as a personal nutrition guide or prescriptive manual. Instead, its aim is to synthesize and interpret findings from peer-reviewed research studies, presenting an overview of the current scientific landscape without injecting the author's personal opinions or anecdotal experiences. The intention is to empower readers with evidence-based knowledge, rather than to promote a particular dietary philosophy or commercial product.

5.7.1   The Confidence Illusion

The educational paradox is only one aspect of how authority is constructed in today's nutrition and dieting landscape. Increasingly, perceived expertise is established not just through academic credentials or technical knowledge, but through the confident delivery of information itself. This psychological phenomenon (known as the "Confidence Heuristic") refers to our tendency to equate self-assuredness with genuine expertise, regardless of the speaker's actual qualifications or the accuracy of their claims. Pulford et al. (2018) demonstrated that people are systematically more likely to trust, follow, and defer to individuals who state their opinions with unwavering certainty, independent of the correctness of those opinions [334].

Unlike the slow accumulation of genuine nutrition expertise through years of study and professional practice, confidence can be instantly manufactured. Charismatic influencers and self-styled nutrition "truth-tellers" exploit this bias by adopting the visual and rhetorical signals of authority: assertive tone, seamless delivery, and the cadence of certainty. Their statements may be speculative, misleading, or even outright invented, yet audiences (often primed by distrust in mainstream dietary guidelines or scientific institutions) are drawn to confidence as a more compelling signal of truth than any peer-reviewed citation or formal degree. In this environment, the authority conferred by confidence can easily surpass that of genuine expertise.

A vivid example of this can be found in the world of nutrition and dieting influencers. For instance, individuals with no formal training in nutrition may confidently assert that seed oils are "toxic" or that a single food group is the root of all modern health problems. Their delivery, brimming with certainty and urgency, persuades large audiences to adopt extreme dietary regimens or reject well-established nutrition science. Podcast hosts, interviewers, and followers may accept these claims at face value, swayed more by the influencer's confidence than by the actual scientific validity of their arguments. As Pulford et al. note, people "are disproportionately influenced by the opinion of the most confident group member, even if this opinion is incorrect" [334]; a bias that is readily exploited by those seeking clicks, sales, or notoriety in the crowded nutrition marketplace.

The consequences are significant. Figures with little to no background in nutritional science can amass huge followings simply by declaring, with conviction, that certain foods are "dangerous" or that mainstream dietary advice is a "scam." The confidence heuristic transforms speculation, cherry-picked anecdotes, and outright misinformation into viral certainties. Meanwhile, qualified experts (who are typically more measured and transparent about scientific uncertainty) may be dismissed or overshadowed for appearing less resolute.

This dynamic further exacerbates the educational paradox. Not only can advanced degrees and technical jargon be misused to lend credibility to unsupported dietary claims, but sheer confidence alone can outcompete true expertise and robust evidence. In today's algorithm-driven attention economy, the loudest and most self-assured voices often dominate the public conversation on diet and nutrition, while genuine experts (trained in humility and scientific caution) are sidelined or dismissed as "uncertain." This creates a distorted hierarchy of authority, incentivizing performance over substance and making it even harder for the public to distinguish credible nutrition information from persuasive, but unfounded, claims.

> **Key Point:** In the nutrition and dieting arena, confidence often masquerades as expertise; making it easy for charismatic but unqualified voices to overshadow evidence-based guidance. Recognizing this "confidence heuristic" is vital for the public to critically assess dietary claims and prioritize genuine expertise over persuasive delivery.

5.7.2 Resilience Limited by Personality Traits

Resilience and advanced education, while valuable, do not necessarily ensure sound scientific judgment or rational decision-making in nutrition and health. In fact, certain personality traits can significantly undermine one's ability to use intelligence and education effectively [335]. One such trait is intellectual arrogance or hyper-confidence: a state in which individuals become so convinced of their intellectual superiority that they disregard evidence or expertise outside their own discipline.

This cognitive pitfall is particularly problematic in nutrition discourse. Higher intelligence can actually exacerbate these issues; individuals with greater cognitive ability are often especially skilled at rationalizing their pre-existing dietary beliefs and constructing sophisticated arguments to justify incorrect or unsubstantiated conclusions [336]. For instance, a medical doctor with a background in an unrelated specialty might confidently dismiss established nutritional science, promoting fad diets or making unsupported claims about food additives while ignoring the expertise of registered dietitians and nutrition researchers.

This dynamic highlights the vital role of intellectual humility in scientific rigor. Intelligence alone is not enough; the willingness to question one's own assumptions and remain open to correction is essential for genuine progress in nutrition science. When intelligence is coupled with traits like narcissism or excessive self-confidence, it can paradoxically lead to greater resistance to new evidence and a tendency to dismiss legitimate expertise outside one's own narrow field.

Ultimately, in nutrition as in all scientific domains, personality traits such as humility and openness to learning are often more important than raw intellect when it comes to evaluating evidence and making sound decisions. Without these qualities, even the most educated and intelligent individuals can fall prey to dogmatism, misinformation, and ultimately, poor health outcomes.

> **Key Point:** In nutrition science, intellectual humility and openness to new evidence are just as important as intelligence or education. Without these qualities, even highly educated individuals can fall into the trap of rationalizing personal biases, spreading misinformation, and making poor dietary decisions; highlighting the need for self-awareness and critical reflection in evaluating nutrition claims [335, 336].

5.8   Monetization of Manufactured Authority

The paradox of education, as discussed previously, is most starkly illustrated by the rise of credentialed contrarians, i.e., individuals who leverage legitimate scientific or medical credentials to challenge and undermine the very consensus their training once supported. This phenomenon is increasingly visible in the world of nutrition and diet, where professionals with advanced degrees use their authority to question or even reject established dietary science.

Rather than being outliers, these figures have become archetypes for a new era of misinformation entrepreneurs. Their credentials and backgrounds serve as shields against criticism and as magnets for audiences seeking "insider" knowledge. Yet their public statements about nutrition, diet, and health are often at odds with the scientific consensus: they may be based on cherry-picked data, outdated models, or outright fabrications. For instance, claims that certain food additives or ingredients are universally toxic, or that there is no meaningful difference between naturally occurring nutrients and synthetic additives, often contradict extensive toxicological and nutritional research.

This misuse of credentials is central to the current "economy of distrust." The authority these individuals wield does not stem from ongoing research or peer-reviewed contributions in nutrition science, but from their ability to position themselves as whistleblowers; purportedly exposing the "hidden dangers" of mainstream dietary guidelines, food safety regulations, or conventional nutrition advice. Their arguments, often delivered with the rhetorical polish of seasoned professionals, are designed to sow doubt rather than provide clarity. They may invoke outdated nutritional models to claim, for example, that certain fats or carbohydrates "skew" metabolism in ways contradicted by current research, or misrepresent the history of nutrition-related diseases by attributing positive health trends to unrelated factors, while ignoring clear epidemiological evidence that supports established interventions.

The economic motivations for this manufactured authority are clear. The alternative nutrition and wellness industries are deeply intertwined with the spread of dietary skepticism. Many credentialed contrarians transition from clinical or academic roles to careers built on book sales, online platforms, speaking engagements, and the promotion of unregulated supplements and "natural" alternatives to mainstream dietary guidance. This business model thrives on the very mistrust it helps to create, transforming skepticism into a self-perpetuating cycle of fear, misinformation, and profit.

A defining feature of this dynamic is the selective use of data and the deliberate oversimplification or distortion of scientific nuance. Isolated or preliminary findings are cherry-picked, stripped of their context, and presented as definitive proof of harm or benefit, while the broader body of methodologically rigorous research is overlooked. For example, a single small-scale study might be cited to claim that a common nutritional intervention causes harm, without disclosing that large-scale clinical trials show no such effect or that the cited study's limitations are substantial and unaddressed. Similarly, debunked or retracted nutrition studies are sometimes defended or cited without acknowledging their methodological or ethical flaws, or the overwhelming evidence that refutes their conclusions.

The consequences of credentialed contrarianism are not abstract. When public trust in evidence-based nutrition erodes, communities may abandon proven dietary interventions, leading to the resurgence of preventable deficiencies and related health conditions. The burden of these outcomes falls not on the influencers profiting from fear and confusion, but on the most vulnerable individuals; such as children, the elderly, and those already at risk for poor nutrition.

In this way, the economy of nutrition contrarianism is both a symptom and a driver of the broader "fear economy." It monetizes distrust, exploits gaps in public understanding, and weaponizes the very symbols of expertise that once anchored confidence in nutrition science [334–337].

> **Key Point:** Credentialed contrarians exploit their authority to cast doubt on established nutrition science, often prioritizing attention and profit over evidence. This dynamic erodes public trust, undermines effective dietary interventions, and amplifies the spread of misinformation; demonstrating the urgent need to critically assess claims based on the quality of evidence, not just the credentials of the messenger.

5.9   Systemic Incentives

The internet not only accelerates the spread of nutrition misrepresentations but also provides fertile ground for individuals and organizations to exploit them for personal, commercial, or ideological gain. In this digital landscape, the long-standing issue of scientific distortion has become more pervasive and urgent, demanding vigilant attention from nutrition scientists, communicators, and the public alike.

5.9.1   Clickbait Economics

A prime example is the popularization of turmeric as a so-called "superfood." Online platforms have amplified turmeric's reputation, touting it as a cure-all for everything from inflammation to cancer, thanks to the curcumin compound it contains [338]. However, much of the supporting research uses curcumin concentrations far higher than what is realistically consumed in a typical diet. Despite these limitations, social media posts and online articles often promote turmeric as a simple, everyday solution for serious health issues, creating unrealistic expectations and fueling a booming market for supplements. This oversimplification not only distorts the science but also diverts attention from the broader context of balanced nutrition and healthy living.

The internet has fostered an environment where sensational headlines and viral content consistently take precedence over accuracy. Simplified or distorted nutrition claims can now reach millions within hours, often stripped of critical nuance. For example, headlines such as "Chocolate Boosts IQ," "Ozempic as a Weight Loss Miracle," or "Turmeric Cures Cancer" are more likely to go viral than nuanced discussions of the evidence and study limitations. As a result, the public is inundated with oversimplified or outright misleading nutrition information, making it increasingly difficult to separate fact from fiction.

These examples underscore a central challenge in the popularization of nutrition science: the relentless drive for clicks, views, and shares. Today's media landscape rewards sensationalism, not accuracy. A headline like "Chocolate Makes You Smarter!" will inevitably attract more attention than a careful explanation of a study's methodological caveats. Similarly, bold but unsupported claims about miracle diets, weight loss drugs, or "detox" foods are more likely to spread than evidence-based, nuanced guidance.

This attention economy creates a feedback loop in which nutrition science is routinely distorted as it's adapted for mass consumption. Writers, influencers, and editors are pressured to craft stories that grab attention, often at the expense of scientific rigor. The end result is a public flooded with misleading or incomplete dietary information; a dynamic that fuels confusion, undermines trust in legitimate nutrition research, and can ultimately lead to poor dietary choices and health outcomes.

Figure 5.4 visualizes the ongoing tension in nutrition communication between evidence-based rigor and sensationalized dietary narratives. The horizontal axis spans from well-substantiated nutrition science on the left to viral clickbait diet headlines on the right. As one moves toward the right, the reliability of information tends to decline while the likelihood of sensationalism increases. The vertical axis contrasts highly technical, often inaccessible expert language at the bottom with sensationalized or oversimplified nutrition messages at the top. The dashed curve depicts the erosion of scientific rigor as sensationalism grows, while the wavy curve demonstrates how exaggerated or misleading nutrition claims can overshadow credible dietary information. The star marker at the center represents the

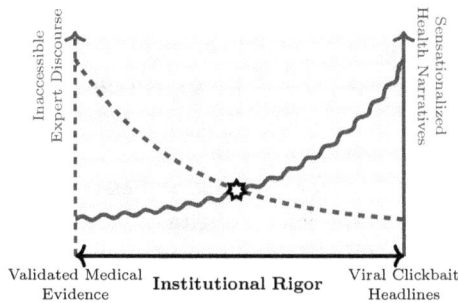

Figure 5.4: This figure depicts the tension in nutrition communication between evidence-based accuracy (shown by the lighter gray curve) and sensationalized dietary claims (shown by the darker gray curve). The horizontal axis spans from thoroughly validated nutrition science on the left to viral, clickbait diet headlines on the right. The vertical axis contrasts highly technical, inaccessible expert language with oversimplified or sensationalized nutrition narratives. The star marker highlights the ideal intersection between these extremes, representing the vital point where scientific credibility and public engagement align; fostering informed dietary choices and greater trust in nutrition information.

ideal balance point (where scientific accuracy and public engagement are harmonized) emphasizing the need for nutrition communication that is both trustworthy and accessible.

> **Key Point:** In the digital age, nutrition information is often distorted as it's adapted for mass appeal; favoring sensational headlines over scientific accuracy. Striking a balance between accessibility and evidence-based rigor is essential for empowering healthier choices and maintaining public trust in nutrition science.

5.9.2   From Simplification to Exploitation

The viral rise of Ozempic as a weight loss "miracle" is a clear example of how oversimplified nutrition narratives can drive misguided behaviors. Originally approved for type 2 diabetes, Ozempic attracted widespread attention for its appetite-suppressing effects, quickly becoming a trending topic on social media as an easy solution for weight loss. This rapid popularization often omitted crucial medical caveats: its off-label use for weight management requires physician oversight due to risks such as nausea, vomiting [339], the potential for long-term dependency [340], and unresolved concerns about a possible association with thyroid cancer (though human evidence remains inconclusive) [341]. The media's focus on Ozempic has also overshadowed the significance of holistic and sustainable lifestyle interventions in addressing weight-related health issues [342]. Such oversimplified portrayals not only mislead the public but also foster misuse, with some individuals seeking the drug without proper medical guidance.

The allure of simple fixes for complex problems is a recurring theme in nutrition misinformation, especially as digital platforms amplify old patterns of distortion. During the COVID-19 pandemic, for instance, hydroxychloroquine was prematurely celebrated as a "miracle cure." Early, inconclusive studies lost context across viral headlines, with methodological flaws, dosage risks, and even study retractions frequently ignored. Social media fueled a narrative framing the drug as a "common sense" remedy suppressed by authorities. This oversimplification distracted from established preventive measures like masking and led to shortages for patients with lupus who relied on the drug. Meanwhile, opportunists profited by selling unregulated versions online and painting skepticism as elitist obstructionism.

Nutrition communication is especially vulnerable to these cycles of distortion. The popular belief that "natural means safe" distorts public understanding, fueling the popularity of herbal supplements and essential oils as alternatives to evidence-based health strategies. Wellness influencers exploit this misconception, promoting products like colloidal silver (linked to irreversible organ damage) as immune boosters, while "detox" regimens misconstrue the liver's natural function as a deficiency needing correction. These narratives often prey on communities marginalized by healthcare inequities, channeling billions into unproven supplements as distrust in mainstream institutions grows.

This cycle repeats in modern forms: the 5G-cancer conspiracy theory, for example, misrepresented decades of research on non-ionizing radiation, recasting it as a "toxic" pandemic culprit. Viral posts blurred the distinction between radio waves and DNA-damaging ionizing radiation, favoring emotional, simplistic narratives over scientific nuance. The impact extended beyond misinformation: cell tower arson in the UK and Netherlands disrupted vital services, while "EMF detox" marketers profited from fear by selling pseudoscientific products. This mirrors historic tendencies to seek scapegoats in times of crisis, now supercharged by algorithms that convert fear into profit.

Nutrition myths often follow this template: anti-vaccine rhetoric persists through anecdotes and cherry-picked data, ignoring population-level evidence. Discredited claims (such as temporal correlations between vaccination and autism) are monetized through "detox" protocols and unproven supplements, echoing the era of Radithor tonics but with global reach through platforms like Instagram. Fad diets, such as the alkaline regimen, oversimplify metabolism by reducing it to pH levels, labeling foods like citrus as "acidic dangers" and selling cookbooks as liberation from "Big Pharma" dogma.

Systemic incentives further entrench misinformation: social media algorithms reward sensational content over nuanced science, and anti-science influencers profit handsomely through affiliate sales of detox products. Cognitive biases, including a preference for simple villains over multifactorial explanations, allow bad actors to reframe obesity or chronic disease as issues solvable by quick fixes rather than systemic change. Communities with histories of institutional betrayal (from the opioid crisis to colonial medical abuses) are especially vulnerable to narratives that conflate skepticism with critical thinking, perpetuating cycles of distrust and exploitation.

Without greater transparency and accountability in nutrition communication, oversimplification will remain both a symptom and a driver of public health crises; ensuring that each new generation faces its own wave of misleading "miracle cures" and dietary fads, often outpacing the spread of sound, evidence-based remedies.

> **Key Point:** Oversimplified and sensationalized nutrition narratives can fuel misinformation, risky health behaviors, and public distrust. Combating these cycles requires transparent, evidence-based communication that addresses complexity and resists the lure of quick fixes or viral trends.

5.9.3 The Quantity Manipulation Paradox

A common strategy among nutrition conspiracy theorists is to inundate skeptics with a barrage of claims (ranging from the dangers of seed oils to supposed "toxins" in fortified foods) relying on the mistaken belief that the sheer quantity of allegations lends legitimacy to their arguments. This tactic leverages a cognitive bias where individuals conflate the number of assertions with the strength of the underlying evidence. Those promoting dietary conspiracies often presume that, even if most of their claims are debunked, the sheer volume will compel others to accept that at least a few must be valid. This rhetorical maneuver aims to create a facade of credibility based on repetition and abundance, rather than on scientifically substantiated evidence.

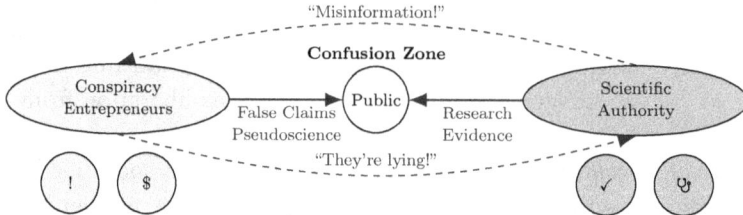

Figure 5.5: This figure depicts the adversarial information dynamics that undermine public trust in nutrition science. At the center is the Public, positioned within a "Confusion Zone" where they are bombarded by contradictory dietary claims. On the left, Conspiracy Entrepreneurs spread pseudoscientific nutrition narratives (such as "False Claims" and "Misinformation"), while on the right, Scientific Authorities present evidence-based dietary guidance. Bidirectional dashed arrows represent mutual accusations ("They're lying!" vs. "Misinformation!"), a pattern known as "flooding the zone." This overwhelming influx of conflicting nutrition information overwhelms critical thinking, especially among individuals lacking the time or expertise to assess each claim. The resulting general distrust of all sources—including reputable scientific channels—deepens contrarian attitudes. Ultimately, the saturation of misleading information ("Information Flood Zone") serves the interests of bad-faith actors, as widespread confusion becomes a powerful driver of dietary conspiracy beliefs and misinformation adoption.

This environment poses significant challenges for the general public; especially those who are either impressionable or simply lack the time and resources to critically evaluate nutrition claims. Many people end up relying on hearsay or the most confidently delivered messages, accepting dietary assertions at face value. On one side are profit-motivated marketers and conspiracy entrepreneurs eager to exploit anxieties around food, nutrition, and health by selling pseudoscientific diets, miracle supplements, or detox regimens. On the other side stand scientists, nutrition experts, and public health authorities attempting to alert consumers to misinformation and protect them from being misled.

When both camps accuse each other of dishonesty and misinformation, the public can become overwhelmed and disillusioned, often concluding that neither side is trustworthy. This "both sides" skepticism, especially when paired with the "flooding the zone" tactics of misinformation purveyors (who inundate audiences with a barrage of conflicting and often false nutrition claims) makes it ever more difficult for individuals to discern credible advice from pseudoscience. The resulting adversarial information landscape further erodes trust in the scientific consensus on nutrition, as visualized in

Figure 5.5.

The paradox in this dynamic is that it turns the scientific method on its head: instead of building carefully on robust evidence for each dietary claim, nutrition conspiracy theorists rely on sheer quantity, flooding the conversation with a multitude of assertions. This tactic overwhelms critical thinking and creates the illusion of credibility through volume rather than substance. The pattern echoes historical responses to crises; such as during the medieval plagues, when a growing collection of witchcraft accusations was mistaken for proof, despite lacking any real causal evidence.

For those committed to evidence-based nutrition, seeking "common ground" becomes futile when the entire argument is built on a foundation of unsubstantiated or fabricated claims. The core challenge for science communicators is to convey why the sheer number of allegations or testimonials in nutrition debates is no substitute for rigorously verified data; a task made even more difficult by social media algorithms that reward the spread of quantity over quality in nutrition information.

> **Key Point:** In nutrition debates, the sheer volume of claims does not equal credibility; only rigorous, well-substantiated evidence should guide dietary decisions. Recognizing and resisting "flooding the zone" tactics is vital for maintaining trust in science and safeguarding public health against misinformation.

5.9.4   The Search Engine Paradox

Conventional wisdom suggests that encouraging people to "do their own research" about nutrition online is an effective defense against misinformation; a digital-age extension of critical thinking. Yet, one of the most troubling paradoxes of the information era is that searching online to verify nutrition or diet claims can actually increase belief in falsehoods. In a landmark series of experiments, Aslett et al. examined whether searching for information online helps users discern truth from misinformation across a range of topics, including health and COVID-19 [343]. The results were striking: participants prompted to use search engines were significantly more likely to rate misleading or outright false articles as credible, with this effect persisting months later and applying across diverse domains.

The underlying mechanism is known as the "data void": when search engines lack high-quality, authoritative coverage for newly viral or fringe nutrition topics, they tend to serve up low-quality or ideologically slanted sources. Instead of encountering corrective information, users are presented with a chorus of unreliable sites, creating a false sense of corroboration. This phenomenon is especially pronounced among those with limited digital literacy or those who search using sensationalist queries (such as pasting clickbait headlines or dubious URLs directly into the search bar). Notably, when search results were dominated by reputable sources, belief in misinformation did not increase; highlighting that the problem lies not with searching itself, but with the quality of the content surfaced and the strategies employed by users.

The paradox deepens: online searching can also increase belief in true stories, but this effect is most pronounced for true claims found on low-quality sites, with little impact on trust in reputable, mainstream nutrition resources. In essence, search engines act as force multipliers for whatever dominates the results page; be it credible science or clickbait. This mirrors the real-world effect of media literacy campaigns that encourage the public to "Google it" without equipping them with the skills to evaluate credibility or navigate information voids. As with the "quantity manipulation paradox" in nutrition, well-meaning advice can inadvertently fuel confusion when users are left to sift through environments already saturated with misleading content.

The implications for nutrition communication and public health are profound. Campaigns that encourage independent searching as a stand-alone solution risk backfiring, especially in fast-moving controversies where reputable sources are slow to respond and opportunistic actors have already optimized their content for visibility. The "do your own research" mantra, instead of protecting individuals, can function as a delivery system for the very misinformation that conspiracy theorists and predatory supplement marketers thrive on. As Aslett et al. argue, media literacy education must go beyond platitudes, grounding recommendations in empirical evidence and teaching people not just to search, but how to search effectively. For search engines, this means a responsibility to prioritize high-quality nutrition information, highlight authoritative fact-checks, and resist manipulation by coordinated campaigns.

A promising development is the introduction of AI-generated summaries at the top of search results. These summaries, which draw from the most authoritative and up-to-date sources, are less vulnerable to the pitfalls of data voids and SEO-driven manipulation. When implemented well, AI overviews can present users with balanced, evidence-based nutrition guidance; steering them away from low-quality content that might otherwise dominate results. In this way, AI-generated summaries represent an encouraging step toward solving the paradox identified by Aslett et al.: by offering reliable syntheses before users encounter a sea of unreliable links, search platforms can disrupt the cycle that has made "do your own research" a vector for confusion. While these tools are not perfect and require ongoing oversight, their emergence signals a positive evolution; one with real potential to restore trust and empower informed inquiry in nutrition science.

Ultimately, the search engine paradox is emblematic of the broader "economy of distrust": a system in which the very mechanisms designed to foster inquiry can, if left unchecked, deepen confusion and erode public trust in nutrition science. Only by reimagining both digital literacy efforts and the architectures of search platforms can this cycle be broken. The rise of AI-generated summaries, coupled with robust transparency and continual improvement, offers hope that future generations will be equipped to navigate nutrition information with greater clarity and confidence.

> **Key Point:** Simply searching online is not enough to protect against nutrition misinformation; without digital literacy and high-quality sources, "doing your own research" can actually reinforce false beliefs. Empowering the public with critical evaluation skills and improving search platform reliability are essential steps for fostering genuine understanding and trust in nutrition science [343].

5.10 AI in Nutrition

As the digital landscape evolves, AI tools such as large language models and chatbots are rapidly becoming popular sources of nutrition advice. When compared to the thousands of misleading, monetized social media videos (many designed to sow distrust in evidence-based guidance in order to boost clicks and supplement sales) AI can, in principle, offer a more balanced, up-to-date, and accessible entry point for those seeking dietary information.

However, the real-world risks of relying solely on AI for nutrition guidance have already become apparent. In a striking case published in the *Annals of Internal Medicine: Clinical Cases* [344], a 60-year-old man was hospitalized for three weeks after he replaced ordinary table salt (sodium chloride) with sodium bromide following a consultation with ChatGPT. The man, with no prior psychiatric history, became confused and paranoid, eventually requiring involuntary psychiatric care after experiencing hallucinations and paranoia. Laboratory tests, combined with consultation from poison control, revealed bromide toxicity; a rare but serious syndrome largely unseen since the days when bromide compounds were common in over-the-counter medicines. It was only after his recovery that the patient revealed he had been conducting a "personal experiment" to eliminate table salt from his diet after reading about its potential dangers and, seeking alternatives, consulted ChatGPT for advice. He subsequently followed the chatbot's suggestion of sodium bromide as a substitute for around three months, unaware of the risks.

The physicians investigating the case noted that, while they did not have the actual chat logs, when they themselves asked ChatGPT-3.5 about possible chloride substitutes, the bot suggested bromide without a clear warning or follow-up questions that a medical professional would have offered. Though the AI mentioned that "context matters," it failed to provide a specific health warning or inquire about the reason for the substitution; an omission with serious consequences in this case. This episode underscores several important lessons for the use of AI in nutrition:

○ **AI is only as safe as its design and oversight.** While it can summarize research and guidelines, AI lacks true clinical judgment and may not provide context-specific warnings or follow-up questions necessary for individual safety.

○ **Critical thinking and verification remain essential.** Any nutrition advice from an AI (especially involving substitutes, supplements, or major dietary changes) should be verified with a qualified health professional.

○ **Responsible AI development is crucial.** Developers must build in robust safety checks, clear disclaimers, and mechanisms to encourage users to seek professional advice for medical or dietary concerns.

Encouragingly, AI can be a valuable tool for navigating the overwhelming landscape of nutrition information, especially compared to profit-driven, sensationalist online content. Unlike many influencers and viral videos that prioritize engagement over accuracy, AI systems (when properly designed) can help users sift through claims and identify evidence-based guidance. One practical way to harness this potential is to use AI as a fact-checker: individuals can upload a video, paste a link, or share the text of a nutrition claim and ask the AI to assess its validity. The AI can then cross-reference the content with reputable scientific literature, highlight inconsistencies, and provide context about the strength of the evidence behind a claim. This approach empowers users to critically evaluate the flood of nutrition advice they encounter online, reducing reliance on anecdotal or monetized misinformation.

However, as the bromide poisoning case demonstrates, AI advice should always be viewed as a starting point; not a substitute for professional consultation or personal research grounded in credible sources. While AI can help flag questionable claims and direct users toward more reliable information, it is not infallible and may occasionally miss context, nuance, or emerging risks. For significant dietary changes or medical concerns, users should always consult with qualified health professionals. In this way, AI becomes a useful ally in the fight against nutrition misinformation; provided it is used thoughtfully, with a healthy dose of skepticism and a commitment to verifying recommendations before acting on them.

AI-powered nutrition tools hold enormous potential for democratizing access to evidence-based guidance, but users must approach them with healthy skepticism and always double-check recommendations; especially those that seem unusual or counterintuitive. As AI continues to improve, its greatest value may be in helping users cut through the noise of online misinformation, provided it is used thoughtfully and never as a replacement for true clinical expertise.

> **Key Point:** AI can be a powerful ally for evidence-based nu-
> trition guidance (far better than relying on monetized misinfor-
> mation) but only when paired with critical thinking, personal
> responsibility, and professional oversight. Always verify AI-
> generated dietary advice before making significant changes to
> your health routine.

Bibliography

[1] Dr. Shilpa Balaji Asegaonkar. Insights into role of adipose tissue as endocrine organ. *International Journal of Diabetes Research*, 1(1):01–04, January 2019. doi:10.33545/26648822.2019.v1.i1a.1.

[2] Alexandros Vegiopoulos, Maria Rohm, and Stephan Herzig. Adipose tissue: between the extremes. *The EMBO Journal*, 36(14):1999–2017, June 2017. doi:10.15252/embj.201696206.

[3] F. Lonnqvist. The obese (OB) gene and its product leptin–a new route toward obesity treatment in man? *QJM*, 89(5):327–332, May 1996. doi:10.1093/qjmed/89.5.327.

[4] Julie A. Chowen and Jesús Argente. Leptin and the brain. *HMBCI*, 7(2):351–360, August 2011. doi:10.1515/hmbci.2011.113.

[5] Jeffrey S Flier and Eleftheria Maratos-Flier. Obesity and the hypothalamus: Novel peptides for new pathways. *Cell*, 92(4):437–440, February 1998. doi:10.1016/s0092-8674(00)80937-x.

[6] Milen Hristov. Leptin signaling in the hypothalamus: Cellular insights and therapeutic perspectives in obesity. *Endocrines*, 6(3):42, August 2025. doi:10.3390/endocrines6030042.

[7] Jiarui Liu, Futing Lai, Yujia Hou, and Ruimao Zheng. Leptin signaling and leptin resistance. *Medical Review*, 2(4):363–384, August 2022. doi:10.1515/mr-2022-0017.

[8] Cheryl D. Fryar, Jacqueline D. Wright, Mark S. Eberhardt, and Bruce A. Dye. Trends in nutrient intakes and chronic health conditions among mexican-american adults, a 25-year profile: United states, 1982–2006. Technical Report 50, National Center for Health Statistics, Hyattsville, MD, 2012. URL: https://www.cdc.gov/nchs/data/nhsr/nhsr050.pdf.

[9] Edward Archer, Gregory A. Hand, and Steven N. Blair. Validity of u.s. nutritional surveillance: National health and nutrition examination survey caloric energy intake data, 1971–2010. *PLoS ONE*, 8(10):e76632, 2013. doi:10.1371/journal.pone.0076632.

[10] Marjorie R. Freedman, Victor L. Fulgoni, and Harris R. Lieberman. Temporal changes in micronutrient intake among united states adults, NHANES 2003 through 2018: A cross-sectional study. *The American Journal of Clinical Nutrition*, 119(6):1309–1320, 2024. doi:10.1016/j.ajcnut.2024.02.007.

[11] Alexandra E. Cowan, Janet A. Tooze, Jaime J. Gahche, Heather A. Eicher-Miller, Patricia M. Guenther, Johanna T. Dwyer, Nancy Potischman, Anindya Bhadra, Raymond J. Carroll, and Regan L. Bailey. Trends in overall and micronutrient-containing dietary supplement use in us adults and children, NHANES 2007–2018. *The Journal of Nutrition*, 152(12):2789–2801, 2022. doi:10.1093/jn/nxac168.

[12] Herman Pontzer, Yosuke Yamada, Hiroyuki Sagayama, Philip N. Ainslie, Lene F. Andersen, Liam J. Anderson, Lenore Arab, Issaad Baddou, Kweku Bedu-Addo, Ellen E. Blaak, Stephane Blanc, Alberto G. Bonomi, Carlijn V.C. Bouten, Pascal Bovet, Maciej S. Buchowski, Nancy F. Butte, Stefan G. Camps, Graeme L. Close, Jamie A. Cooper, Richard Cooper, et al. Daily energy expenditure through the human life course. *Science*, 373(6556):808–812, 2021. doi:10.1126/science.abe5017.

[13] Kay Nguo, Helen Truby, and Judi Porter. Total energy expenditure in healthy ambulatory older adults aged ≥ 80 years: A doubly labelled water study. *Annals of Nutrition and Metabolism*, 79(2):263–273, 2023. doi:10.1159/000528872.

[14] John R. Speakman, Jasper M. A. de Jong, Srishti Sinha, Klaas R. Westerterp, Yosuke Yamada, et al. Total daily energy expenditure has declined over the last 3 decades due to declining basal expenditure not reduced activity expenditure. *Nature Metabolism*, 5(4):579–588, 2023. doi:10.1038/s42255-023-00782-2.

[15] Marie-Pierre St-Onge and Dympna Gallagher. Body composition changes with aging: The cause or the result of alterations in metabolic rate and macronutrient oxidation? *Nutrition*, 26(2):152–155, February 2010. doi:10.1016/j.nut.2009.07.004.

[16] John R. Speakman and Colin Selman. Physical activity and resting metabolic rate. *Proceedings of the Nutrition Society*, 62(3):621–634, 2003. doi:10.1079/PNS2003282.

[17] Allyson K. Palmer and Michael D. Jensen. Metabolic changes in aging humans: current evidence and therapeutic strategies. *The Journal of Clinical Investigation*, 132(16):e158451, 2022. doi:10.1172/JCI158451.

[18] Xiaotao Shen, Chuchu Wang, Xin Zhou, Wenyu Zhou, Daniel Hornburg, Si Wu, and Michael P. Snyder. Nonlinear dynamics of multi-omics profiles during human aging. *Nature Aging*, 4:1619–1634, 2024. doi:10.1038/s43587-024-00692-2.

[19] Paul S. MacLean, Audrey Bergouignan, Marc-Andre Cornier, and Matthew R. Jackman. Biology's response to dieting: the impetus for weight regain. *American Journal of Physiology-Regulatory, Integrative and Comparative Physiology*, 301(3):R581–R600, September 2011. doi:10.1152/ajpregu.00755.2010.

[20] Sanjay R Patel and Frank B Hu. Short sleep duration and weight gain: a systematic review. *Obesity*, 16(3):643–653, 2008. doi:10.1038/oby.2007.118.

[21] Shahrad Taheri, Ling Lin, Diane Austin, Terry Young, and Emmanuel Mignot. Short sleep duration is associated with reduced leptin, elevated ghrelin, and increased body mass index. *PLoS Medicine*, 1(3):e62, 2004. doi:10.1371/journal.pmed.0010062.

[22] Karine Spiegel, Esra Tasali, Plamen Penev, and Eve Van Cauter. Sleep curtailment in healthy young men is associated with decreased leptin levels, elevated ghrelin levels, and increased hunger and appetite. *Annals of Internal Medicine*, 141(11):846–850, 2004. doi:10.7326/0003-4819-141-11-200412070-00008.

[23] Esra Tasali, Rachel Leproult, and Karine Spiegel. Reduced sleep duration or quality: relationships with insulin resistance and type 2 diabetes. *Progress in Cardiovascular Diseases*, 51(5):381–391, 2009. doi:10.1016/j.pcad.2008.10.002.

[24] Arlet V Nedeltcheva, Jennifer M Kilkus, Jacqueline Imperial, Kristen Kasza, Dale A Schoeller, and Plamen D Penev. Sleep curtailment is accompanied by increased intake of calories from snacks. *American Journal of Clinical Nutrition*, 89:126–133, 2009. doi:10.3945/ajcn.2008.26574.

[25] Ariana M. Chao, Ania M. Jastreboff, Marney A. White, Carlos M. Grilo, and Rajita Sinha. Stress, cortisol, and other appetite-related hormones: Prospective prediction of 6-month changes in food cravings and weight. *Obesity*, 25(4):713–720, 2017. doi:10.1002/oby.21790.

[26] Susanne Kuckuck, Eline S. van der Valk, Robin Lengton, Julius März, Manon H.J. Hillegers, Brenda W.J.H. Penninx, Maryam Kavousi, Mariëtte R. Boon, Sjoerd A.A. van den Berg, and Elisabeth F.C. van Rossum. Long-term hair cortisone and perceived stress are associated with long-term hedonic eating tendencies in patients with obesity. *Psychoneuroendocrinology*, 171:107224, 2025. doi:10.1016/j.psyneuen.2024.107224.

[27] Rose Seoyoung Chang, Hilâl Cerit, Taryn Hye, E. Leighton Durham, Harlyn Aizley, Sarah Boukezzi, Florina Haimovici, Jill M. Goldstein, Daniel G. Dillon, Diego A. Pizzagalli, and Laura M. Holsen. Stress-induced alterations in hpa-axis reactivity and mesolimbic reward activation in individuals with emotional eating. *Appetite*, 168:105707, 2022. doi:10.1016/j.appet.2021.105707.

[28] Elena-Gabriela Strete, Mădălina-Gabriela Cincu, and Andreea Sălcudean. Disordered eating behaviors, perceived stress and insomnia during academic exams: A study among university students. *Medicina*, 61(7):1226, 2025. doi: 10.3390/medicina61071226.

[29] Merve Pehlivan, Neslişah Denkçi, Reyhan Pehlivan, Muhammet Ali undefinedakır, and Yeliz Mercan. The relationship between body image and nutritional behaviors in adult individuals. *PLOS ONE*, 20(3):e0320408, March 2025. doi:10.1371/journal.pone.0320408.

[30] Hamid Sharif-Nia, Erika Sivarajan Froelicher, Ozkan Gorgulu, Jason W. Osborne, Aleksandra Błachnio, Azadeh Rezazadeh Fazeli, Amir Hossein Goudarzian, and Omolhoda Kaveh. The relationship among positive body image, body esteem, and eating attitude in iranian population. *Frontiers in Psychology*, 15, February 2024. doi:10.3389/fpsyg.2024.1304555.

[31] Jessica A. Malloy, Hugo Kazenbroot-Phillips, and Rajshri Roy. Associations between body image, eating behaviors, and diet quality among young women in new zealand: The role of social media. *Nutrients*, 16(20):3517, October 2024. doi:10.3390/nu16203517.

[32] Alina Zaharia and Iulia Gonța. The healthy eating movement on social media and its psychological effects on body image. *Frontiers in Nutrition*, 11, December 2024. doi:10.3389/fnut.2024.1474729.

[33] Kim Rounsefell, Simone Gibson, Siân McLean, Merran Blair, Annika Molenaar, Linda Brennan, Helen Truby, and Tracy A. McCaffrey. Social media, body image

and food choices in healthy young adults: A mixed methods systematic review. *Nutrition & Dietetics*, 77(1):19–40, October 2019. `doi:10.1111/1747-0080.12581`.

[34] Ana Maria Jiménez-García, Natalia Arias, Elena Picazo Hontanaya, Ana Sanz, and Olivia García-Velasco. Impact of body-positive social media content on body image perception. *Journal of Eating Disorders*, 13(1), July 2025. `doi:10.1186/s40337-025-01286-y`.

[35] Basmah Suliman Salman Alburkani, Fatimah M. Yousef, Arwa Arab, and Afnan. A. Qutub. The impact of social media and family attitudes on the body image and eating patterns of male and female students. *Middle East Current Psychiatry*, 31(1), October 2024. `doi:10.1186/s43045-024-00474-x`.

[36] John D. Eastwood, Alexandra Frischen, Mark J. Fenske, and Daniel Smilek. The unengaged mind: Defining boredom in terms of attention. *Perspectives on Psychological Science*, 7(5):482–495, September 2012. `doi:10.1177/1745691612456044`.

[37] Yael K. Goldberg, John D. Eastwood, Jennifer LaGuardia, and James Danckert. Boredom: An emotional experience distinct from apathy, anhedonia, or depression. *Journal of Social and Clinical Psychology*, 30(6):647–666, June 2011. `doi:10.1521/jscp.2011.30.6.647`.

[38] Andrew B. Moynihan, Wijnand A. P. van Tilburg, Eric R. Igou, Arnaud Wisman, Alan E. Donnelly, and Jessie B. Mulcaire. Eaten up by boredom: consuming food to escape awareness of the bored self. *Frontiers in Psychology*, 6, April 2015. `doi:10.3389/fpsyg.2015.00369`.

[39] Amanda C Crockett, Samantha K Myhre, and Paul D Rokke. Boredom proneness and emotion regulation predict emotional eating. *Journal of Health Psychology*, 20(5):670–680, April 2015. `doi:10.1177/1359105315573439`.

[40] Remco C. Havermans, Linda Vancleef, Antonis Kalamatianos, and Chantal Nederkoorn. Eating and inflicting pain out of boredom. *Appetite*, 85:52–57, February 2015. `doi:10.1016/j.appet.2014.11.007`.

[41] Federica Grant, Maria Luisa Scalvedi, Umberto Scognamiglio, Aida Turrini, and Laura Rossi. Eating habits during the covid-19 lockdown in italy: The nutritional and lifestyle side effects of the pandemic. *Nutrients*, 13(7):2279, June 2021. `doi:10.3390/nu13072279`.

[42] Hanna Konttinen, Tatjana van Strien, Satu Männistö, Pekka Jousilahti, and Ari Haukkala. Depression, emotional eating and long-term weight changes: a population-based prospective study. *International Journal of Behavioral Nutrition and Physical Activity*, 16(1), March 2019. `doi:10.1186/s12966-019-0791-8`.

[43] Camille Lassale, G. David Batty, Amaria Baghdadli, Felice Jacka, Almudena Sánchez-Villegas, Mika Kivimäki, and Tasnime Akbaraly. Healthy dietary indices and risk of depressive outcomes: a systematic review and meta-analysis of observational studies. *Molecular Psychiatry*, 24(7):965–986, September 2018. `doi:10.1038/s41380-018-0237-8`.

[44] Fatemeh Maleki Sedgi, Jalal Hejazi, Reza Derakhshi, Ghazal Baghdadi, Melinaz Zarmakhi, Mana Hamidi, Kamyar Mansori, Mohsen Dadashi, and Mehran Rahimlou. Investigation of the relationship between food preferences and depression symptoms among undergraduate medical students: a cross-sectional study. *Frontiers in Nutrition*, 12, March 2025. `doi:10.3389/fnut.2025.1519726`.

[45] Yichun Zhang, Mingzhu Zhang, Qihua Guan, Pengcheng Liu, Wangxin Zhang, Song Yang, Hualei Dong, and Haifeng Hou. Dietary intake and five types of mental disorders: a bidirectional mendelian randomization study. *BMC Psychiatry*, 25(1), July 2025. doi:10.1186/s12888-025-07100-y.

[46] Alia J. Crum and Ellen J. Langer. Mind-set matters: Exercise and the placebo effect. *Psychological Science*, 18(2):165–171, February 2007. doi:10.1111/j.1467-9280.2007.01867.x.

[47] Chanmo Park, Francesco Pagnini, Andrew Reece, Deborah Phillips, and Ellen Langer. Blood sugar level follows perceived time rather than actual time in people with type 2 diabetes. *Proceedings of the National Academy of Sciences*, 113(29):8168–8170, July 2016. doi:10.1073/pnas.1603444113.

[48] Chanmo Park, Francesco Pagnini, and Ellen Langer. Glucose metabolism responds to perceived sugar intake more than actual sugar intake. *Scientific Reports*, 10(1), September 2020. doi:10.1038/s41598-020-72501-w.

[49] Alia J. Crum, William R. Corbin, Kelly D. Brownell, and Peter Salovey. Mind over milkshakes: Mindsets, not just nutrients, determine ghrelin response. *Health Psychology*, 30(4):424–429, 2011. doi:10.1037/a0023467.

[50] Peter Aungle and Ellen Langer. Physical healing as a function of perceived time. *Scientific Reports*, 13(1), December 2023. doi:10.1038/s41598-023-50009-3.

[51] Lisa T. Crummett and Riley J. Grosso. Postprandial glycemic response to whole fruit versus blended fruit in healthy, young adults. *Nutrients*, 14(21):4565, October 2022. doi:10.3390/nu14214565.

[52] Rabab Alkutbe, Kathy Redfern, Michael Jarvis, and Gail Rees. Nutrient extraction lowers postprandial glucose response of fruit in adults with obesity as well as healthy weight adults. *Nutrients*, 12(3):766, March 2020. doi:10.3390/nu12030766.

[53] Barbara B. Kahn and Jeffrey S. Flier. Obesity and insulin resistance. *Journal of Clinical Investigation*, 106(4):473–481, August 2000. doi:10.1172/jci10842.

[54] Xiaoxuan Liu, Huimin Zhou, Yixian Liu, Jinhong Li, Huijing Luo, Qian He, Yanv Ren, Xiaofang Zhang, and Zuoliang Dong. Exploring insulin resistance and pancreatic function in individuals with overweight and obesity: Insights from ogtts and irts. *Diabetes Research and Clinical Practice*, 219:111972, January 2025. doi:10.1016/j.diabres.2024.111972.

[55] Himan Mohamed-Mohamed, Teresa Pardo-Moreno, Margarita Jimenez-Palomares, Bibiana Perez-Ardanaz, Encarnación M. Sánchez-Lara, Maria D. Vazquez-Lara, Mario de La Mata-Fernandez, Victoria García-Morales, and Juan José Ramos-Rodríguez. Impaired glucose tolerance and altered body composition in obese young adults: A case–control study. *Biomedicines*, 13(7):1569, June 2025. doi:10.3390/biomedicines13071569.

[56] Julie E. Flood-Obbagy and Barbara J. Rolls. The effect of fruit in different forms on energy intake and satiety at a meal. *Appetite*, 52(2):416–422, April 2009. doi:10.1016/j.appet.2008.12.001.

[57] Gwen Chodur, Jody Randolph, Allison Bardagjy, and Francene Steinberg. The ingestion of whole fruit compared to fruit juice differentially modulates glucose, insulin and inflammatory pathways. *Current Developments in Nutrition*, 4:nzaa058_009, June 2020. doi:10.1093/cdn/nzaa058_009.

[58] Nuria Grigelmo-Miguel and Olga Martin-Belloso. Characterization of dietary fiber from orange juice extraction. *Food Research International*, 31(5):355–361, June 1998. doi:10.1016/s0963-9969(98)00087-8.

[59] Christine Schmucker, Angelika Eisele-Metzger, Joerg J Meerpohl, Cornelius Lehane, Daniela Kuellenberg de Gaudry, Szimonetta Lohner, and Lukas Schwingshackl. Effects of a gluten-reduced or gluten-free diet for the primary prevention of cardiovascular disease. *Cochrane Database of Systematic Reviews*, 2022(2), February 2022. doi:10.1002/14651858.cd013556.pub2.

[60] Glenn A. Gaesser and Siddhartha S. Angadi. Gluten-free diet: Imprudent dietary advice for the general population? *Journal of the Academy of Nutrition and Dietetics*, 112(9):1330–1333, September 2012. doi:10.1016/j.jand.2012.06.009.

[61] Norelle R. Reilly. The gluten-free diet: Recognizing fact, fiction, and fad. *The Journal of Pediatrics*, 175:206–210, August 2016. doi:10.1016/j.jpeds.2016.04.014.

[62] Mari C. W. Myhrstad, Marlene Slydahl, Monica Hellmann, Lisa Garnweidner-Holme, Knut E. A. Lundin, Christine Henriksen, and Vibeke H. Telle-Hansen. Nutritional quality and costs of gluten-free products: a case-control study of food products on the norwegian marked. *Food & Nutrition Research*, 65, March 2021. doi:10.29219/fnr.v65.6121.

[63] Jessica R Biesiekierski, Evan D Newnham, Peter M Irving, Jacqueline S Barrett, Melissa Haines, James D Doecke, Susan J Shepherd, Jane G Muir, and Peter R Gibson. Gluten causes gastrointestinal symptoms in subjects without celiac disease: A double-blind randomized placebo-controlled trial. *American Journal of Gastroenterology*, 106(3):508–514, March 2011. doi:10.1038/ajg.2010.487.

[64] Benjamin Niland and Brooks D Cash. Health benefits and adverse effects of a gluten-free diet in non-celiac disease patients. *Gastroenterology & Hepatology*, 14(2):82–91, 2018.

[65] Jessica R. Biesiekierski, Simone L. Peters, Evan D. Newnham, Ourania Rosella, Jane G. Muir, and Peter R. Gibson. No effects of gluten in patients with self-reported non-celiac gluten sensitivity after dietary reduction of fermentable, poorly absorbed, short-chain carbohydrates. *Gastroenterology*, 145(2):320–328.e3, August 2013. doi:10.1053/j.gastro.2013.04.051.

[66] Jessica R. Biesiekierski, Jane G. Muir, and Peter R. Gibson. Is gluten a cause of gastrointestinal symptoms in people without celiac disease? *Current Allergy and Asthma Reports*, 13(6):631–638, September 2013. doi:10.1007/s11882-013-0386-4.

[67] Benjamin Lebwohl, Yin Cao, Geng Zong, Frank B Hu, Peter H R Green, Alfred I Neugut, Eric B Rimm, Laura Sampson, Lauren W Dougherty, Edward Giovannucci, Walter C Willett, Qi Sun, and Andrew T Chan. Long term gluten consumption in adults without celiac disease and risk of coronary heart disease: prospective cohort study. *BMJ*, page j1892, May 2017. doi:10.1136/bmj.j1892.

[68] Guy H. Johnson and Kevin Fritsche. Effect of dietary linoleic acid on markers of inflammation in healthy persons: A systematic review of randomized controlled trials. *Journal of the Academy of Nutrition and Dietetics*, 112(7):1029–1041.e15, July 2012. doi:10.1016/j.jand.2012.03.029.

[69] Jun Li, Marta Guasch-Ferré, Yanping Li, and Frank B Hu. Dietary intake and biomarkers of linoleic acid and mortality: systematic review and meta-analysis of prospective cohort studies. *The American Journal of Clinical Nutrition*, 112(1):150–167, July 2020. doi:10.1093/ajcn/nqz349.

[70] Kristina S. Petersen, Mark Messina, and Brent Flickinger. Health implications of linoleic acid and seed oil intake. *Nutrition Today*, March 2025. doi:10.1097/nt. 0000000000000746.

[71] Maryam S. Farvid, Ming Ding, An Pan, Qi Sun, Stephanie E. Chiuve, Lyn M. Steffen, Walter C. Willett, and Frank B. Hu. Dietary linoleic acid and risk of coronary heart disease: A systematic review and meta-analysis of prospective cohort studies. *Circulation*, 130(18):1568–1578, October 2014. doi:10.1161/circulationaha.114.010236.

[72] Matti Marklund, Jason H.Y. Wu, Fumiaki Imamura, Liana C. Del Gobbo, Amanda Fretts, Janette de Goede, Peilin Shi, Nathan Tintle, Maria Wennberg, Stella Aslibekyan, Tzu-An Chen, Marcia C. de Oliveira Otto, Yoichiro Hirakawa, Helle Højmark Eriksen, Janine Kröger, Federica Laguzzi, Maria Lankinen, Rachel A. Murphy, Kiesha Prem, Cécilia Samieri, et al. Biomarkers of dietary omega-6 fatty acids and incident cardiovascular disease and mortality: An individual-level pooled analysis of 30 cohort studies. *Circulation*, 139(21):2422–2436, May 2019. doi:10.1161/circulationaha.118.038908.

[73] Yu Zhang, Katia S. Chadaideh, Yanping Li, Yuhan Li, Xiao Gu, Yuxi Liu, Marta Guasch-Ferré, Eric B. Rimm, Frank B. Hu, Walter C. Willett, Meir J. Stampfer, and Dong D. Wang. Butter and plant-based oils intake and mortality. *JAMA Internal Medicine*, 185(5):549, May 2025. doi:10.1001/jamainternmed.2025.0205.

[74] Kevin L. Fritsche. Too much linoleic acid promotes inflammation—doesn't it? *Prostaglandins, Leukotrienes and Essential Fatty Acids*, 79(3–5):173–175, September 2008. doi:10.1016/j.plefa.2008.09.019.

[75] Kristina S. Petersen, Kevin C. Maki, Philip C. Calder, Martha A. Belury, Mark Messina, Carol F. Kirkpatrick, and William S. Harris. Perspective on the health effects of unsaturated fatty acids and commonly consumed plant oils high in unsaturated fat. *British Journal of Nutrition*, 132(8):1039–1050, October 2024. doi:10.1017/s0007114524002459.

[76] Marco Witkowski, Ina Nemet, Hassan Alamri, Jennifer Wilcox, Nilaksh Gupta, Nisreen Nimer, Arash Haghikia, Xinmin S. Li, Yuping Wu, Prasenjit Prasad Saha, Ilja Demuth, Maximilian König, Elisabeth Steinhagen-Thiessen, Tomas Cajka, Oliver Fiehn, Ulf Landmesser, W. H. Wilson Tang, and Stanley L. Hazen. The artificial sweetener erythritol and cardiovascular event risk. *Nature Medicine*, 29(3):710–718, February 2023. doi:10.1038/s41591-023-02223-9.

[77] Tagreed A. Mazi and Kimber L. Stanhope. Erythritol: An in-depth discussion of its potential to be a beneficial dietary component. *Nutrients*, 15(1):204, January 2023. doi:10.3390/nu15010204.

[78] F.R.J. Bornet, A. Blayo, F. Dauchy, and G. Slama. Plasma and urine kinetics of erythritol after oral ingestion by healthy humans. *Regulatory Toxicology and Pharmacology*, 24(2):S280–S285, October 1996. doi:10.1006/rtph.1996.0109.

[79] Matthew D. Ritchey, Hilary K. Wall, Mary G. George, and Janet S. Wright. Us trends in premature heart disease mortality over the past 50 years: Where do we go from here? *Trends in Cardiovascular Medicine*, 30(6):364–374, August 2020. doi:10.1016/j.tcm.2019.09.005.

[80] George A. Mensah, Gina S. Wei, Paul D. Sorlie, Lawrence J. Fine, Yves Rosenberg, Peter G. Kaufmann, Michael E. Mussolino, Lucy L. Hsu, Ebyan Addou, Michael M. Engelgau, and David Gordon. Decline in cardiovascular mortality: Possible causes and implications. *Circulation Research*, 120(2):366–380, January 2017. doi:10.1161/circresaha.116.309115.

[81] Hannah K. Weir, Robert N. Anderson, Sallyann M. Coleman King, Ashwini Soman, Trevor D. Thompson, Yuling Hong, Bjorn Moller, and Steven Leadbetter. Heart disease and cancer deaths — trends and projections in the united states, 1969–2020. *Preventing Chronic Disease*, 13, November 2016. doi:10.5888/pcd13.160211.

[82] Seth S. Martin, Aaron W. Aday, Zaid I. Almarzooq, Cheryl A.M. Anderson, Pankaj Arora, Christy L. Avery, Carissa M. Baker-Smith, Bethany Barone Gibbs, Andrea Z. Beaton, Amelia K. Boehme, Yvonne Commodore-Mensah, Maria E. Currie, Mitchell S.V. Elkind, Kelly R. Evenson, Giuliano Generoso, Debra G. Heard, Swapnil Hiremath, Michelle C. Johansen, Rizwan Kalani, Dhruv S. Kazi, et al. 2024 heart disease and stroke statistics: A report of us and global data from the american heart association. *Circulation*, 149(8), February 2024. doi:10.1161/cir.0000000000001209.

[83] Richard E. Goodman. Twenty-eight years of gm food and feed without harm: why not accept them? *GM Crops & Food*, 15(1):40–50, March 2024. doi:10.1080/21645698.2024.2305944.

[84] Alessandro Nicolia, Alberto Manzo, Fabio Veronesi, and Daniele Rosellini. An overview of the last 10 years of genetically engineered crop safety research. *Critical Reviews in Biotechnology*, 34(1):77–88, September 2013. doi:10.3109/07388551.2013.823595.

[85] Xu Hui, Randy Kwaku Amponsah, Samuel Antwi, Patrick Kweku Gbolonyo, Moses Agyemang Ameyaw, Geoffrey Bentum-Micah, and Edward Oppong Adjei. Understanding the societal dilemma of genetically modified food consumption: a stimulus-organism-response investigation. *Frontiers in Sustainable Food Systems*, 8, September 2024. doi:10.3389/fsufs.2024.1364052.

[86] Zexin Chen, Weimin Xu, Yangmei Huang, Xingyi Jin, Jin Deng, Sujuan Zhu, Hui Liu, Shanchun Zhang, and Yunxian Yu. Associations of noniodized salt and thyroid nodule among the chinese population: a large cross-sectional study. *The American Journal of Clinical Nutrition*, 98(3):684–692, September 2013. doi:10.3945/ajcn.112.054353.

[87] Isabela P. Loyola, Mauri Felix de Sousa, Thiago Veiga Jardim, Marcela M. Mendes, Weimar Kunz Sebba Barroso, Ana Luiza Lima Sousa, and Paulo Cesar B. Veiga Jardim. Comparison between the effects of hymalaian salt and common salt intake on urinary sodium and blood pressure in hypertensive individuals. *Arquivos Brasileiros de Cardiologia*, 118(5):875–882, December 2022. doi:10.36660/abc.20210069.

[88] Giuseppe Lisco, Anna De Tullio, Domenico Triggiani, Roberta Zupo, Vito Angelo Giagulli, Giovanni De Pergola, Giuseppina Piazzolla, Edoardo Guastamacchia,

Carlo Sabbà, and Vincenzo Triggiani. Iodine deficiency and iodine prophylaxis: An overview and update. *Nutrients*, 15(4):1004, February 2023. doi:10.3390/nu15041004.

[89] Uffe Ravnskov, David M Diamond, Rokura Hama, Tomohito Hamazaki, Björn Hammarskjöld, Niamh Hynes, Malcolm Kendrick, Peter H Langsjoen, Aseem Malhotra, Luca Mascitelli, Kilmer S McCully, Yoichi Ogushi, Harumi Okuyama, Paul J Rosch, Tore Schersten, Sherif Sultan, and Ralf Sundberg. Lack of an association or an inverse association between low-density-lipoprotein cholesterol and mortality in the elderly: a systematic review. *BMJ Open*, 6(6):e010401, June 2016. doi:10.1136/bmjopen-2015-010401.

[90] Shereif H. Rezkalla and Robert A. Kloner. Low HDL—the challenge. *Clinical Medicine & Research*, 23(2):60–66, August 2025. doi:10.3121/cmr.2025.1970.

[91] William B. Kannel, Thomas R. Dawber, Abraham Kagan, Naomi Revotskie, and Joseph Stokes. Factors of risk in the development of coronary heart disease–six year follow-up experience. the framingham study. *Annals of Internal Medicine*, 55:33–50, 1961. doi:10.7326/0003-4819-55-1-33.

[92] Caroline Andersson, Andrew D. Johnson, Emelia J. Benjamin, Daniel Levy, and Ramachandran S. Vasan. 70-year legacy of the framingham heart study. *Nature Reviews Cardiology*, 16(11):687–698, 2019. doi:10.1038/s41569-019-0202-5.

[93] H. C. Jr. McGill. The relationship of dietary cholesterol to serum cholesterol concentration and to atherosclerosis in man. *American Journal of Clinical Nutrition*, 32(12 Suppl):2664–2702, 1979. doi:10.1093/ajcn/32.12.2664.

[94] Joseph L. Goldstein and Michael S. Brown. The low-density lipoprotein pathway and its relation to atherosclerosis. *Annual Review of Biochemistry*, 46:897–930, 1977. doi:10.1146/annurev.bi.46.070177.004341.

[95] Frank M. Sacks, Marc A. Pfeffer, Lemuel A. Moye, Jean L. Rouleau, John D. Rutherford, Thomas G. Cole, Lisa Brown, J. Wayne Warnica, J. Malcolm O. Arnold, Chuan-Chuan Wun, Barry R. Davis, and Eugene Braunwald. The effect of pravastatin on coronary events after myocardial infarction in patients with average cholesterol levels. *New England Journal of Medicine*, 335(14):1001–1009, October 1996. doi:10.1056/nejm199610033351401.

[96] Christopher P. Cannon, Eugene Braunwald, Charles H. McCabe, and et al. Intensive versus moderate lipid lowering with statins after acute coronary syndromes. *New England Journal of Medicine*, 350(15):1495–1504, 2004. doi:10.1056/NEJMoa040583.

[97] Marc S. Sabatine, Robert P. Giugliano, Anthony C. Keech, and et al. Evolocumab and clinical outcomes in patients with cardiovascular disease. *New England Journal of Medicine*, 376(18):1713–1722, 2017. doi:10.1056/NEJMoa1615664.

[98] Christopher P. Cannon, Michael A. Blazing, Robert P. Giugliano, and et al. Ezetimibe added to statin therapy after acute coronary syndromes. *New England Journal of Medicine*, 372(25):2387–2397, 2015. doi:10.1056/NEJMoa1410489.

[99] Martijn Vergeer, Anke G. Holleboom, John J.P. Kastelein, and Jan A. Kuivenhoven. The HDL hypothesis: does high-density lipoprotein protect from atherosclerosis? *Journal of Lipid Research*, 51(8):2058–2073, 2010. doi:10.1194/jlr.R001610.

[100] G. Howard Rothblat and M. C. Phillips. High-density lipoprotein heterogeneity and function in reverse cholesterol transport. *Current Opinion in Lipidology*, 21(3):229–238, 2010. doi:10.1097/mol.0b013e328338472d.

[101] G. G. Schwartz, P. G. Steg, M. Szarek, et al. Alirocumab and cardiovascular outcomes after acute coronary syndrome. *New England Journal of Medicine*, 379(22):2097–2107, 2018. doi:10.1056/NEJMoa1801174.

[102] G. G. Schwartz, A. G. Olsson, M. Abt, et al. Effects of dalcetrapib in patients with a recent acute coronary syndrome. *New England Journal of Medicine*, 367(22):2089–2099, 2012. doi:10.1056/NEJMoa1206797.

[103] A. M. Lincoff, S. J. Nicholls, J. S. Riesmeyer, et al. Evacetrapib and cardiovascular outcomes in high-risk vascular disease. *New England Journal of Medicine*, 376(20):1933–1942, 2017. doi:10.1056/NEJMoa1609581.

[104] L. Bowman, J. C. Hopewell, K. Wallendszus, et al. Effects of anacetrapib in patients with atherosclerotic vascular disease. *New England Journal of Medicine*, 377(13):1217–1227, 2017. doi:10.1056/NEJMoa1706444.

[105] S. Kodama, S. Tanaka, K. Saito, et al. Effect of aerobic exercise training on serum levels of high-density lipoprotein cholesterol: a meta-analysis. *Archives of Internal Medicine*, 167(10):999–1008, 2007. doi:10.1001/archinte.167.10.999.

[106] C. Couillard, J. P. Després, B. Lamarche, et al. Effects of endurance exercise training on plasma hdl cholesterol levels depend on levels of triglycerides: evidence from men of the health, risk factors, exercise training and genetics (heritage) family study. *Arteriosclerosis, Thrombosis, and Vascular Biology*, 21(7):1226–1232, 2001. doi:10.1161/hq0701.092137.

[107] Antonella Muscella, Erika Stefàno, and Santo Marsigliante. The effects of exercise training on lipid metabolism and coronary heart disease. *American Journal of Physiology-Heart and Circulatory Physiology*, 319(1):H76–H88, July 2020. doi:10.1152/ajpheart.00708.2019.

[108] G. M. Berger. High-density lipoproteins in the prevention of atherosclerotic heart disease. part i. epidemiological and family studies. *South African Medical Journal*, 54(17):689–693, 1978. URL: https://pubmed.ncbi.nlm.nih.gov/217108/.

[109] N. A. Zakai, J. Minnier, M. M. Safford, et al. Race-dependent association of high-density lipoprotein cholesterol levels with incident coronary artery disease. *Journal of the American College of Cardiology*, 80(22):2104–2115, 2022. doi:10.1016/j.jacc.2022.09.027.

[110] Jean-Philippe Drouin-Chartier, Amanda L Schwab, Siyu Chen, Yanping Li, Frank M Sacks, Bernard Rosner, JoAnn E Manson, Walter C Willett, Meir J Stampfer, Frank B Hu, and Shilpa N Bhupathiraju. Egg consumption and risk of type 2 diabetes: findings from 3 large us cohort studies of men and women and a systematic review and meta-analysis of prospective cohort studies. *The American Journal of Clinical Nutrition*, 112(3):619–630, September 2020. doi:10.1093/ajcn/nqaa115.

[111] Alice Wallin, Nita G. Forouhi, Alicja Wolk, and Susanna C. Larsson. Egg consumption and risk of type 2 diabetes: a prospective study and dose–response meta-analysis. *Diabetologia*, 59(6):1204–1213, March 2016. doi:10.1007/s00125-016-3923-6.

[112] Mahshid Dehghan, Andrew Mente, and Salim Yusuf. Eggs and diabetes: 1 daily egg a safe bet? *The American Journal of Clinical Nutrition*, 112(3):503–504, September 2020. doi:10.1093/ajcn/nqaa183.

[113] Nicholas R Fuller, Ian D Caterson, Amanda Sainsbury, Gareth Denyer, Mackenzie Fong, James Gerofi, Katherine Baqleh, Kathryn H Williams, Namson S Lau, and Tania P Markovic. The effect of a high-egg diet on cardiovascular risk factors in people with type 2 diabetes: the diabetes and egg (diabegg) study—a 3-mo randomized controlled trial. *The American Journal of Clinical Nutrition*, 101(4):705–713, April 2015. doi:10.3945/ajcn.114.096925.

[114] Man-Yun Li, Jin-Hua Chen, Chiehfeng Chen, and Yi-No Kang. Association between egg consumption and cholesterol concentration: A systematic review and meta-analysis of randomized controlled trials. *Nutrients*, 12(7):1995, July 2020. doi:10.3390/nu12071995.

[115] Mingyang Song, Teresa T. Fung, Frank B. Hu, Walter C. Willett, Valter D. Longo, Andrew T. Chan, and Edward L. Giovannucci. Association of animal and plant protein intake with all-cause and cause-specific mortality. *JAMA Internal Medicine*, 176(10):1453, October 2016. doi:10.1001/jamainternmed.2016.4182.

[116] K Beyreuther, H K Biesalski, J D Fernstrom, P Grimm, W P Hammes, U Heinemann, O Kempski, P Stehle, H Steinhart, and R Walker. Consensus meeting: monosodium glutamate – an update. *European Journal of Clinical Nutrition*, 61(3):304–313, September 2006. doi:10.1038/sj.ejcn.1602526.

[117] John D. Fernstrom and Miro Smriga. Letter-to-the-editor: Shannon m. et al., 2017. toxicology letters 265 (97). *Toxicology Letters*, 272:101–102, April 2017. doi:10.1016/j.toxlet.2017.03.004.

[118] Johanna Aguachela, Andrea Aldaz, Diego Allo, Esteban Vargas, and Carlos Jácome. Monosodium glutamate in prepared foods: Impact on food and its effect on health. *Journal of Agro-industry Sciences*, 4(1):23–27, March 2022. doi:10.17268/jais.2022.003.

[119] R.A. Kenney. The chinese restaurant syndrome: An anecdote revisited. *Food and Chemical Toxicology*, 24(4):351–354, April 1986. doi:10.1016/0278-6915(86)90014-1.

[120] Rabbani Syed Imam. Genotoxicity of monosodium glutamate: A review on its causes, consequences and prevention. *Indian Journal of Pharmaceutical Education and Research*, 53(4s):s510–s517, November 2019. doi:10.5530/ijper.53.4s.145.

[121] John D. Fernstrom. Monosodium glutamate in the diet does not raise brain glutamate concentrations or disrupt brain functions. *Annals of Nutrition and Metabolism*, 73(Suppl. 5):43–52, 2018. doi:10.1159/000494782.

[122] Selamat Jinap, Parvaneh Hajeb, Roslina Karim, Sarian Norliana, Simayi Yibadatihan, and Razak Abdul-Kadir. Reduction of sodium content in spicy soups using monosodium glutamate. *Food & Nutrition Research*, 60(1):30463, January 2016. doi:10.3402/fnr.v60.30463.

[123] Jeremia Halim, Ali Bouzari, Dan Felder, and Jean-Xavier Guinard. The salt flip: Sensory mitigation of salt (and sodium) reduction with monosodium glutamate (msg) in "better-for-you" foods. *Journal of Food Science*, 85(9):2902–2914, August 2020. doi:10.1111/1750-3841.15354.

[124] Martha Ballesteros, Fabrizio Valenzuela, Alma Robles, Elizabeth Artalejo, David Aguilar, Catherine Andersen, Herlindo Valdez, and Maria Fernandez. One egg per day improves inflammation when compared to an oatmeal-based breakfast without increasing other cardiometabolic risk factors in diabetic patients. *Nutrients*, 7(5):3449–3463, May 2015. doi:10.3390/nu7053449.

[125] Sun Jo Kim, Cheol Woon Jung, Nguyen Hoang Anh, Suk Won Kim, Seongoh Park, Sung Won Kwon, and Seul Ji Lee. Effects of oats (avena sativa l.) on inflammation: A systematic review and meta-analysis of randomized controlled trials. *Frontiers in Nutrition*, 8, August 2021. doi:10.3389/fnut.2021.722866.

[126] Qingtao Hou, Yun Li, Ling Li, Gaiping Cheng, Xin Sun, Sheyu Li, and Haoming Tian. The metabolic effects of oats intake in patients with type 2 diabetes: A systematic review and meta-analysis. *Nutrients*, 7(12):10369–10387, December 2015. doi:10.3390/nu7125536.

[127] Zoltan Ungvari, Mónika Fekete, Péter Varga, Andrea Lehoczki, Gyöngyi Munkácsy, János Tibor Fekete, Giampaolo Bianchini, Alberto Ocana, Annamaria Buda, Anna Ungvari, and Balázs Győrffy. Association between red and processed meat consumption and colorectal cancer risk: a comprehensive meta-analysis of prospective studies. *GeroScience*, 47(3):5123–5140, April 2025. doi:10.1007/s11357-025-01646-1.

[128] Maryam S. Farvid, Elkhansa Sidahmed, Nicholas D. Spence, Kingsly Mante Angua, Bernard A. Rosner, and Junaidah B. Barnett. Consumption of red meat and processed meat and cancer incidence: a systematic review and meta-analysis of prospective studies. *European Journal of Epidemiology*, 36(9):937–951, August 2021. doi:10.1007/s10654-021-00741-9.

[129] Anika Knuppel, Keren Papier, Georgina K Fensom, Paul N Appleby, Julie A Schmidt, Tammy Y N Tong, Ruth C Travis, Timothy J Key, and Aurora Perez-Cornago. Meat intake and cancer risk: prospective analyses in uk biobank. *International Journal of Epidemiology*, 49(5):1540–1552, August 2020. doi:10.1093/ije/dyaa142.

[130] Nour Makarem, Elisa V. Bandera, Joseph M. Nicholson, and Niyati Parekh. Consumption of sugars, sugary foods, and sugary beverages in relation to cancer risk: A systematic review of longitudinal studies. *Annual Review of Nutrition*, 38:17–39, 2018. doi:10.1146/annurev-nutr-082117-051805.

[131] Yin Huang, Zeyu Chen, Bo Chen, Jinze Li, Xiang Yuan, Jin Li, Wen Wang, Tingting Dai, Hongying Chen, Yan Wang, Ruyi Wang, Puze Wang, Jianbing Guo, Qiang Dong, Chengfei Liu, Qiang Wei, Dehong Cao, and Liangren Liu. Dietary sugar consumption and health: umbrella review. *BMJ*, 381:e071609, 2023. doi:10.1136/bmj-2022-071609.

[132] Eloi Chazelas, Bernard Srour, Elisa Desmetz, Emmanuelle Kesse-Guyot, Chantal Julia, Valérie Deschamps, Nathalie Druesne-Pecollo, Pilar Galan, Serge Hercberg, Paule Latino-Martel, Mélanie Deschasaux, and Mathilde Touvier. Sugary drink consumption and risk of cancer: results from nutrinet-santé prospective cohort. *BMJ*, 365:l2408, 2019. doi:10.1136/bmj.l2408.

[133] Fjorida Llaha, Mercedes Gil-Lespinard, Pelin Unal, Izar de Villasante, Jazmín Castañeda, and Raul Zamora-Ros. Consumption of sweet beverages and cancer risk. a systematic review and meta-analysis of observational studies. *Nutrients*, 13(2):516, 2021. doi:10.3390/nu13020516.

[134] Tomoyoshi Aoyagi, Krista P. Terracina, Ali Raza, Hisahiro Matsubara, and Kazuaki Takabe. Cancer cachexia, mechanism and treatment. *World Journal of Gastrointestinal Oncology*, 7(4):17–29, 2015. doi:10.4251/wjgo.v7.i4.17.

[135] Rubén Fernández-Rodríguez, Bruno Bizzozero-Peroni, Valentina Díaz-Go ni, Miriam Garrido-Miguel, Gabriele Bertotti, Alberto Roldán-Ruiz, and Miguel López-Moreno. Plant-based meat alternatives and cardiometabolic health: a systematic review and meta-analysis. *The American Journal of Clinical Nutrition*, 121(2):274–283, 2025. doi:10.1016/j.ajcnut.2024.12.002.

[136] Matthew Nagra, Felicia Tsam, Shaun Ward, and Ehud Ur. Animal vs plant-based meat: A hearty debate. *Canadian Journal of Cardiology*, 40(7):1198–1209, 2024. doi:10.1016/j.cjca.2023.11.005.

[137] Hang Su, Ruijie Liu, Ming Chang, Jianhua Huang, and Xingguo Wang. Dietary linoleic acid intake and blood inflammatory markers: a systematic review and meta-analysis of randomized controlled trials. *Food & Function*, 8(9):3091–3103, 2017. doi:10.1039/c7fo00433h.

[138] Helena Bjermo, David Iggman, Joel Kullberg, Ingrid Dahlman, Lars Johansson, Lena Persson, Johan Berglund, Kari Pulkki, Samar Basu, Matti Uusitupa, Mats Rudling, Peter Arner, Tommy Cederholm, Hakan Ahlstrom, and Ulf Riserus. Effects of n-6 pufas compared with sfas on liver fat, lipoproteins, and inflammation in abdominal obesity: a randomized controlled trial. *The American Journal of Clinical Nutrition*, 95(5):1003–1012, 2012. doi:10.3945/ajcn.111.030114.

[139] Azza O Alawad, Tarig H Merghani, and Mansour A Ballal. Resting metabolic rate in obese diabetic and obese non-diabetic subjects and its relation to glycaemic control. *BMC Research Notes*, 6:382, September 2013. doi:10.1186/1756-0500-6-382.

[140] Nathan Caron, Nicolas Peyrot, Teddy Caderby, Chantal Verkindt, and Georges Dalleau. Energy expenditure in people with diabetes mellitus: A review. *Frontiers in Nutrition*, 3:56, December 2016. doi:10.3389/fnut.2016.00056.

[141] Theresa Drabsch, Christina Holzapfel, Lynne Stecher, Julia Petzold, Thomas Skurk, and Hans Hauner. Associations between c-reactive protein, insulin sensitivity, and resting metabolic rate in adults: A mediator analysis. *Frontiers in Endocrinology*, 9:556, September 2018. doi:10.3389/fendo.2018.00556.

[142] Kevin D Hall, Kong Y Chen, Juen Guo, Yan Y Lam, Rudolph L Leibel, Laurel ES Mayer, Marc L Reitman, Michael Rosenbaum, Steven R Smith, B Timothy Walsh, and Eric Ravussin. Energy expenditure and body composition changes after an isocaloric ketogenic diet in overweight and obese men. *American Journal of Clinical Nutrition*, 104(2):324–333, August 2016. doi:10.3945/ajcn.116.133561.

[143] Christopher D. Gardner, John F. Trepanowski, Liana C. Del Gobbo, Michael E. Hauser, Joseph Rigdon, John P. A. Ioannidis, Manisha Desai, and Abby C. King. Effect of low-fat vs low-carbohydrate diet on 12-month weight loss in overweight adults and the association with genotype pattern or insulin secretion: The dietfits randomized clinical trial. *JAMA*, 319(7):667–679, February 2018. doi:10.1001/jama.2018.0245.

[144] Christopher D. Gardner, Lisa Offringa, Jennifer Hartle, Kris Kapphahn, and Rise Cherin. Weight loss on low-fat vs. low-carb diets by insulin resistance status

among overweight adults & adults with obesity: A randomized pilot trial. *Obesity*, 24(1):79–86, January 2016. doi:10.1002/oby.21331.

[145] Yoh Miyashita, Nobukiyo Koide, Masaki Ohtsuka, Hiroshi Ozaki, Yoshiaki Itoh, Tomokazu Oyama, Takako Uetake, Kiyoko Ariga, and Kohji Shirai. Beneficial effect of low carbohydrate in low calorie diets on visceral fat reduction in type 2 diabetic patients with obesity. *Diabetes Research and Clinical Practice*, 65(3):235–241, September 2004. doi:10.1016/j.diabres.2004.01.008.

[146] Kuo-Chin Huang, Nic Kormas, Katharine Steinbeck, Georgina Loughnan, and Ian D. Caterson. Resting metabolic rate in severely obese diabetic and nondiabetic subjects. *Obesity Research*, 12(5):840–845, May 2004. doi:10.1038/oby.2004.101.

[147] Pedram Pam, Sanaz Asemani, Mohammad Hesam Azizi, and Parmida Jamilian. Chia seed supplementation and inflammatory biomarkers: a systematic review and meta-analysis. *Journal of Nutritional Science*, 13, 2024. doi:10.1017/jns.2024.70.

[148] Mehdi Karimi, Samira Pirzad, Niyousha Shirsalimi, Sajad Ahmadizad, Seyyed Mohammad Hashemi, Shaghayegh Karami, Kimia Kazemi, Erfan Shahir-Roudi, and Anita Aminzadeh. Effects of chia seed (salvia hispanica l.) supplementation on cardiometabolic health in overweight subjects: a systematic review and meta-analysis of rcts. *Nutrition & Metabolism*, 21(1), September 2024. doi:10.1186/s12986-024-00847-3.

[149] Jinhui Zhao, Tim Stockwell, Tim Naimi, Sam Churchill, James Clay, and Adam Sherk. Association between daily alcohol intake and risk of all-cause mortality: A systematic review and meta-analyses. *JAMA Network Open*, 6(3):e236185, March 2023. doi:10.1001/jamanetworkopen.2023.6185.

[150] Susanna C. Larsson, Stephen Burgess, Amy M. Mason, and Karl Michaëlsson. Alcohol consumption and cardiovascular disease: A mendelian randomization study. *Circulation: Genomic and Precision Medicine*, 13(3), June 2020. doi:10.1161/circgen.119.002814.

[151] Vincenzo Bagnardi, Marta Blangiardo, Carlo La Vecchia, and Giovanni Corrao. Alcohol consumption and the risk of cancer: A meta-analysis. *Alcohol Research & Health*, 25(4):263–270, 2001. URL: https://pubmed.ncbi.nlm.nih.gov/11910703/.

[152] Remi Daviet, Gökhan Aydogan, Kanchana Jagannathan, Nathaniel Spilka, Philipp D. Koellinger, Henry R. Kranzler, Gideon Nave, and Reagan R. Wetherill. Associations between alcohol consumption and gray and white matter volumes in the uk biobank. *Nature Communications*, 13(1), March 2022. doi:10.1038/s41467-022-28735-5.

[153] Eivind Molversmyr, Hanne Marie Devle, Carl Fredrik Naess-Andresen, and Dag Ekeberg. Identification and quantification of lipids in wild and farmed atlantic salmon (salmo salar), and salmon feed by GC-MS. *Food Science & Nutrition*, 10:3117–3127, 2022. doi:10.1002/fsn3.2911.

[154] Ida-Johanne Jensen, Karl-Erik Eilertsen, Carina Helen Almli Otnæs, Hanne K. Mæhre, and Edel Oddny Elvevoll. An update on the content of fatty acids, dioxins, pcbs and heavy metals in farmed, escaped and wild atlantic salmon (salmo salar l.) in norway. *Foods*, 9(12):1901, 2020. doi:10.3390/foods9121901.

[155] Jette Jakobsen, Cat Smith, Anette Bysted, and Kevin D. Cashman. Vitamin D in wild and farmed atlantic salmon (salmo salar)—what do we know? *Nutrients*, 11(5):982, 2019. doi:10.3390/nu11050982.

[156] A.-K. Lundebye, E.-J. Lock, J. D. Rasinger, O. J. Nøstbakken, R. Hannisdal, E. Karlsbakk, V. Wennevik, A. S. Madhun, L. Madsen, I. E. Graff, and R. Ørnsrud. Lower levels of persistent organic pollutants, metals and the marine omega 3-fatty acid dha in farmed compared to wild atlantic salmon (salmo salar). *Environmental Research*, 155:49–59, 2017. doi:10.1016/j.envres.2017.01.026.

[157] Artemis P. Simopoulos. The omega-6/omega-3 fatty acid ratio, genetic variation, and cardiovascular disease. *Asia Pacific Journal of Clinical Nutrition*, 17(Suppl 1):131–134, 2008. URL: https://pubmed.ncbi.nlm.nih.gov/18296320/.

[158] Artemis P. Simopoulos. The importance of the ratio of omega-6/omega-3 essential fatty acids. *Biomedicine & Pharmacotherapy*, 56(8):365–379, 2002. doi:10.1016/S0753-3322(02)00253-6.

[159] Yunhe Hong, Nicholas Birse, Brian Quinn, Yicong Li, Wenyang Jia, Saskia van Ruth, and Christopher T. Elliott. MALDI-ToF MS and chemometric analysis as a tool for identifying wild and farmed salmon. *Food Chemistry*, 432:137279, 2024. doi:10.1016/j.foodchem.2023.137279.

[160] Amin Mousavi Khaneghah, Yadolah Fakhri, Leili Abdi, Carolina Fernanda Sengling Cebin Coppa, Larissa Tuanny Franco, and Carlos Augusto Fernandes de Oliveira. The concentration and prevalence of ochratoxin a in coffee and coffee-based products: A global systematic review, meta-analysis and meta-regression. *Fungal Biology*, 123(8):611–617, 2019. doi:10.1016/j.funbio.2019.05.012.

[161] Mehmet Kanbay, Dimitrie Siriopol, Sidar Copur, Laura Tapoi, Laura Benchea, Masanari Kuwabara, Patrick Rossignol, Alberto Ortiz, Adrian Covic, and Baris Afsar. Effect of coffee consumption on renal outcome: A systematic review and meta-analysis of clinical studies. *Journal of Renal Nutrition*, 31(1):5–20, 2021. doi:10.1053/j.jrn.2020.08.004.

[162] Yunping Zhou, Changwei Tian, and Chongqi Jia. A dose–response meta-analysis of coffee consumption and bladder cancer. *Preventive Medicine*, 55(1):14–22, 2012. doi:10.1016/j.ypmed.2012.04.020.

[163] Michael F. Mendoza, Ralf Martz Sulague, Therese Posas-Mendoza, and Carl J. Lavie. Impact of coffee consumption on cardiovascular health. *Ochsner Journal*, 23(2):152–158, 2023. doi:10.31486/toj.22.0073.

[164] Ruth Bonita, Robert Beaglehole, and Tord Kjellström. *Basic Epidemiology*. World Health Organization, Geneva, 2nd edition, 2006. URL: http://whqlibdoc.who.int/publications/2006/9241547073_eng.pdf.

[165] E Strandhagen and D S Thelle. Filtered coffee raises serum cholesterol: results from a controlled study. *European Journal of Clinical Nutrition*, 57(9):1164–1168, August 2003. doi:10.1038/sj.ejcn.1601668.

[166] Asne Lirhus Svatun, Maja-Lisa Løchen, Dag Steinar Thelle, and Tom Wilsgaard. Association between espresso coffee and serum total cholesterol: the tromsø study 2015–2016. *Open Heart*, 9(1):e001946, April 2022. doi:10.1136/openhrt-2021-001946.

[167] ÅB Lirhus Svatun. *Coffee and Cholesterol - Impact of Brewing Methods: From the seventh survey of the Tromsø Study in 2015-2016*. PhD thesis, UiT The Arctic University of Norway, 2021.

[168] AMLETO D'AMICIS, CRISTINA SCACCINI, GIANNI TOMASSI, MICHELE ANACLERIO, RODOLFO STORNELLI, and ALBERTO BERNINI. Italian style brewed coffee: Effect on serum cholesterol in young men. *International Journal of Epidemiology*, 25(3):513–520, 1996. doi:10.1093/ije/25.3.513.

[169] Sara Grioni, Claudia Agnoli, Sabina Sieri, Valeria Pala, Fulvio Ricceri, Giovanna Masala, Calogero Saieva, Salvatore Panico, Amalia Mattiello, Paolo Chiodini, Rosario Tumino, Graziella Frasca, Licia Iacoviello, Amalia de Curtis, Paolo Vineis, and Vittorio Krogh. Espresso coffee consumption and risk of coronary heart disease in a large italian cohort. *PLOS ONE*, 10(5):e0126550, May 2015. doi: 10.1371/journal.pone.0126550.

[170] Mery Yovana Rendón, Maria Brígida dos Santos Scholz, and Neura Bragagnolo. Physical characteristics of the paper filter and low cafestol content filter coffee brews. *Food Research International*, 108:280–285, June 2018. doi:10.1016/j.foodres.2018.03.041.

[171] L Cai, D Ma, Y Zhang, Z Liu, and P Wang. The effect of coffee consumption on serum lipids: a meta-analysis of randomized controlled trials. *European Journal of Clinical Nutrition*, 66(8):872–877, June 2012. doi:10.1038/ejcn.2012.68.

[172] Marcin Bara'nski, Dominika 'Srednicka Tober, Nikolaos Volakakis, Chris Seal, Roy Sanderson, Gavin B Stewart, Charles Benbrook, Bruno Biavati, Emilia Markellou, Charilaos Giotis, et al. Higher antioxidant and lower cadmium concentrations and lower incidence of pesticide residues in organically grown crops: a systematic literature review and meta-analyses. *British Journal of Nutrition*, 112(5):794–811, 2014.

[173] Vanessa Vigar, Stephen Myers, Christopher Oliver, Jacinta Arellano, Shelley Robinson, and Carlo Leifert. A systematic review of organic versus conventional food consumption: is there a measurable benefit on human health? *Nutrients*, 12(1):7, 2020.

[174] Martina Boschiero, Valeria De Laurentiis, Carla Caldeira, and Serenella Sala. Comparison of organic and conventional cropping systems: A systematic review of life cycle assessment studies. *Environmental Impact Assessment Review*, 102:107187, 2023.

[175] Olivia M Smith, Abigail L Cohen, Cassandra J Rieser, Alexandra G Davis, Joseph M Taylor, Adekunle W Adesanya, Matthew S Jones, Amanda R Meier, John P Reganold, Robert J Orpet, et al. Organic farming provides reliable environmental benefits but increases variability in crop yields: a global meta-analysis. *Frontiers in Sustainable Food Systems*, 3:82, 2019.

[176] Amir Hadi, Makan Pourmasoumi, Ameneh Najafgholizadeh, Cain C. T. Clark, and Ahmad Esmaillzadeh. The effect of apple cider vinegar on lipid profiles and glycemic parameters: a systematic review and meta-analysis of randomized clinical trials. *BMC Complementary Medicine and Therapies*, 21(1), June 2021. doi:10.1186/s12906-021-03351-w.

[177] Donya Arjmandfard, Mehrdad Behzadi, Zahra Sohrabi, and Mohsen Moham-
madi Sartang. Effects of apple cider vinegar on glycemic control and insulin
sensitivity in patients with type 2 diabetes: A grade-assessed systematic review
and dose-response meta-analysis of controlled clinical trials. *Frontiers in Nutrition*,
12, January 2025. doi:10.3389/fnut.2025.1528383.

[178] Solaleh Sadat Khezri, Atoosa Saidpour, Nima Hosseinzadeh, and Zohreh Amiri.
Beneficial effects of apple cider vinegar on weight management, visceral adiposity
index and lipid profile in overweight or obese subjects receiving restricted calorie
diet: A randomized clinical trial. *Journal of Functional Foods*, 43:95–102, April
2018. doi:10.1016/j.jff.2018.02.003.

[179] Tomoo KONDO, Mikiya KISHI, Takashi FUSHIMI, Shinobu UGAJIN, and
Takayuki KAGA. Vinegar intake reduces body weight, body fat mass, and serum
triglyceride levels in obese japanese subjects. *Bioscience, Biotechnology, and
Biochemistry*, 73(8):1837–1843, August 2009. doi:10.1271/bbb.90231.

[180] M. C. E. Lomer. Review article: the aetiology, diagnosis, mechanisms and clinical evidence for food intolerance. *Alimentary Pharmacology & Therapeutics*,
41(3):262–275, December 2014. doi:10.1111/apt.13041.

[181] A. Muraro, G. Roberts, M. Worm, M. B. Bilò, K. Brockow, M. Fernández Rivas,
A. F. Santos, Z. Q. Zolkipli, A. Bellou, K. Beyer, C. Bindslev-Jensen, V. Cardona,
A. T. Clark, P. Demoly, A. E. J. Dubois, A. DunnGalvin, P. Eigenmann, S. Halken,
L. Harada, G. Lack, M. Jutel, B. Niggemann, F. Ruëff, F. Timmermans, B. J. Vlieg-
Boerstra, T. Werfel, S. Dhami, S. Panesar, C. A. Akdis, and A. Sheikh. Anaphylaxis:
guidelines from the european academy of allergy and clinical immunology. *Allergy*,
69(8):1026–1045, June 2014. doi:10.1111/all.12437.

[182] Rhoda Sheryl Kagan. Food allergy: an overview. *Environmental Health Perspectives*,
111(2):223–225, February 2003. doi:10.1289/ehp.5702.

[183] Hugh A. Sampson, Seema Aceves, S. Allan Bock, John James, Stacie Jones, David
Lang, Kari Nadeau, Anna Nowak-Wegrzyn, John Oppenheimer, Tamara T. Perry,
Christopher Randolph, Scott H. Sicherer, Ronald A. Simon, Brian P. Vickery,
Robert Wood, Hugh A. Sampson, Christopher Randolph, David Bernstein, Joann
Blessing-Moore, David Khan, David Lang, Richard Nicklas, John Oppenheimer,
Jay Portnoy, Christopher Randolph, Diane Schuller, Sheldon Spector, Stephen A.
Tilles, Dana Wallace, Hugh A. Sampson, Seema Aceves, S. Allan Bock, John
James, Stacie Jones, David Lang, Kari Nadeau, Anna Nowak-Wegrzyn, John
Oppenheimer, Tamara T. Perry, Christopher Randolph, Scott H. Sicherer, Ronald A.
Simon, Brian P. Vickery, and Robert Wood. Food allergy: A practice parameter
update—2014. *Journal of Allergy and Clinical Immunology*, 134(5):1016–1025.e43,
November 2014. doi:10.1016/j.jaci.2014.05.013.

[184] Steven O. Stapel, R. Asero, B. K. Ballmer-Weber, E. F. Knol, S. Strobel, S. Vieths,
and J. Kleine-Tebbe. Testing for IgG4 against foods is not recommended as a
diagnostic tool: EAACI task force report. *Allergy*, 63(7):793–796, June 2008.
doi:10.1111/j.1398-9995.2008.01705.x.

[185] Jacek Gocki and Zbigniew Bartuzi. Role of immunoglobulin G antibodies in
diagnosis of food allergy. *Advances in Dermatology and Allergology*, 4:253–256,
2016. doi:10.5114/ada.2016.61600.

[186] Guohua Zheng, Pingting Qiu, Rui Xia, Huiying Lin, Bingzhao Ye, Jing Tao, and Lidian Chen. Effect of aerobic exercise on inflammatory markers in healthy middle-aged and older adults: A systematic review and meta-analysis of randomized controlled trials. *Frontiers in Aging Neuroscience*, 11, April 2019. doi:10.3389/fnagi.2019.00098.

[187] Dagfinn Aune, Edward Giovannucci, Paolo Boffetta, Lars T Fadnes, NaNa Keum, Teresa Norat, Darren C Greenwood, Elio Riboli, Lars J Vatten, and Serena Tonstad. Fruit and vegetable intake and the risk of cardiovascular disease, total cancer and all-cause mortality—a systematic review and dose-response meta-analysis of prospective studies. *International Journal of Epidemiology*, 46(3):1029–1056, February 2017. doi:10.1093/ije/dyw319.

[188] Lee Hooper, Nicole Martin, Oluseyi F Jimoh, Christian Kirk, Eve Foster, and Asmaa S Abdelhamid. Reduction in saturated fat intake for cardiovascular disease. *Cochrane Database of Systematic Reviews*, 2020(8), August 2020. doi:10.1002/14651858.cd011737.pub3.

[189] Michael E. Makover, Michael D. Shapiro, and Peter P. Toth. There is urgent need to treat atherosclerotic cardiovascular disease risk earlier, more intensively, and with greater precision: A review of current practice and recommendations for improved effectiveness. *American Journal of Preventive Cardiology*, 12:100371, December 2022. doi:10.1016/j.ajpc.2022.100371.

[190] Tanja Kongerslev Thorning, Hanne Christine Bertram, Jean-Philippe Bonjour, Lisette de Groot, Didier Dupont, Emma Feeney, Richard Ipsen, Jean Michel Lecerf, Alan Mackie, Michelle C McKinley, Marie-Caroline Michalski, Didier Rémond, Ulf Risérus, Sabita S Soedamah-Muthu, Tine Tholstrup, Connie Weaver, Arne Astrup, and Ian Givens. Whole dairy matrix or single nutrients in assessment of health effects: current evidence and knowledge gaps ,. *The American Journal of Clinical Nutrition*, 105(5):1033–1045, May 2017. doi:10.3945/ajcn.116.151548.

[191] José Miguel Aguilera. The food matrix: implications in processing, nutrition and health. *Critical Reviews in Food Science and Nutrition*, 59(22):3612–3629, September 2018. doi:10.1080/10408398.2018.1502743.

[192] Edoardo Capuano and Anja E.M. Janssen. Food matrix and macronutrient digestion. *Annual Review of Food Science and Technology*, 12(1):193–212, March 2021. doi:10.1146/annurev-food-032519-051646.

[193] Blerina Shkembi and Thom Huppertz. Calcium absorption from food products: Food matrix effects. *Nutrients*, 14(1):180, December 2021. doi:10.3390/nu14010180.

[194] Jack A. Yanovski, Susan Z. Yanovski, Kara N. Sovik, Tuc T. Nguyen, Patrick M. O'Neil, and Nancy G. Sebring. A prospective study of holiday weight gain. *New England Journal of Medicine*, 342(12):861–867, 2000. doi:10.1056/NEJM200003233421206.

[195] Rolando G. Díaz-Zavala, María F. Castro-Cantú, Mauro E. Valencia, Gerardo Álvarez Hernández, Michelle M. Haby, and Julián Esparza-Romero. Effect of the holiday season on weight gain: A narrative review. *Journal of Obesity*, 2017:1–13, 2017. doi:10.1155/2017/2085136.

[196] David F. Williamson, Henry S. Kahn, Paul L. Remington, and Richard F. Anda. The 10-year incidence of overweight and major weight gain in us adults. *Archives of Internal Medicine*, 150:665–672, 1990. doi:10.1001/archinte.1990.00390150135026.

[197] Graham A. Colditz, Walter C. Willett, Meir J. Stampfer, Susan J. London, Michael R. Segal, and Frank E. Speizer. Patterns of weight change and their relation to diet in a cohort of healthy women. *American Journal of Clinical Nutrition*, 51:1100–1105, 1990. doi:10.1093/ajcn/51.6.1100.

[198] David F. Williamson. Descriptive epidemiology of body weight and weight change in u.s. adults. *Annals of Internal Medicine*, 119(7 Pt 2):646–649, 1993. doi:10.7326/0003-4819-119-7_part_2-199310011-00004.

[199] Robert W. Jeffery and Simone A. French. Preventing weight gain in adults: design, methods and one year results from the pound of prevention study. *International Journal of Obesity and Related Metabolic Disorders*, 21(5):457–464, 1997. doi:10.1038/sj.ijo.0800431.

[200] I. Andersson and S. Rossner. The christmas factor in obesity therapy. *International Journal of Obesity and Related Metabolic Disorders*, 16(12):1013–1015, 1992. URL: http://europepmc.org/abstract/MED/1335971.

[201] G S Andersen, A J Stunkard, T I A Sørensen, L Petersen, and B L Heitmann. Night eating and weight change in middle-aged men and women. *International Journal of Obesity*, 28(10):1338–1343, July 2004. doi:10.1038/sj.ijo.0802731.

[202] Maija B. Bruzas and Kelly C. Allison. A review of the relationship between night eating syndrome and body mass index. *Current Obesity Reports*, 8(2):145–155, February 2019. doi:10.1007/s13679-019-00331-7.

[203] Barbara J. Rolls, Julia A. Ello-Martin, and Beth Carlton Tohill. What can intervention studies tell us about the relationship between fruit and vegetable consumption and weight management? *Nutrition Reviews*, 62(1):1–17, 2004. doi:10.1301/nr.2004.jan.1-17.

[204] Candida J. Rebello, Carol E. O'Neil, and Frank L. Greenway. Dietary fiber and satiety: the effects of oats on satiety. *Nutrition Reviews*, 74(2):131–147, 2016. doi:10.1093/nutrit/nuv063.

[205] A. M. Uhe, G. R. Collier, and K. O'Dea. A comparison of the effects of beef, chicken and fish protein on satiety and amino acid profiles in lean male subjects. *Journal of Nutrition*, 122(3):467–472, 1992. doi:10.1093/jn/122.3.467.

[206] S. H. A. Holt, J. C. Brand Miller, P. Petocz, and E. Farmakalidis. A satiety index of common foods. *European Journal of Clinical Nutrition*, 49:675–690, 1995. URL: https://pubmed.ncbi.nlm.nih.gov/7498104/.

[207] Slavko Komarnytsky, Alison Cook, and Ilya Raskin. Potato protease inhibitors inhibit food intake and increase circulating cholecystokinin levels by a trypsin-dependent mechanism. *International Journal of Obesity (London)*, 35(2):236–243, 2011. doi:10.1038/ijo.2010.192.

[208] E.A. Wambogo, N. Ansai, A. Terry, C. Fryar, and C. Ogden. Dairy, meat, seafood, and plant sources of saturated fat: United states, ages two years and over, 2017–2020. *The Journal of Nutrition*, 153:2689–2698, 2023. doi:10.1016/j.tjnut.2023.06.040.

[209] F. Dayrit. Rethinking saturated fat. *Preprint*, 2023. https://doi.org/10.22541/au.167630583.31924897/v1.

[210] B. Lamarche, A. Astrup, R.H. Eckel, E. Feeney, I. Givens, R.M. Krauss, P. Legrand, R. Micha, M.-C. Michalski, S. Soedamah-Muthu, Q. Sun, and F.J. Kok. Regular-fat and low-fat dairy foods and cardiovascular diseases: perspectives for future dietary recommendations. *The American Journal of Clinical Nutrition*, 121:956–964, 2025. doi:10.1016/j.ajcnut.2025.03.009.

[211] J.-P. Drouin-Chartier, J.A. Côté, M.-E. Labonté, D. Brassard, M. Tessier-Grenier, S. Desroches, P. Couture, and B. Lamarche. Comprehensive review of the impact of dairy foods and dairy fat on cardiometabolic risk. *Advances in Nutrition*, 7:1041–1051, 2016. doi:10.3945/an.115.011619.

[212] Dariush Mozaffarian, Renata Micha, and Sarah Wallace. Effects on coronary heart disease of increasing polyunsaturated fat in place of saturated fat: a systematic review and meta-analysis of randomized controlled trials. *PLoS Medicine*, 7(3):e1000252, 2010. doi:10.1371/journal.pmed.1000252.

[213] Lukas Schwingshackl and Georg Hoffmann. Monounsaturated fatty acids, olive oil and health status: a systematic review and meta-analysis of cohort studies. *Lipids in Health and Disease*, 13:154, 2014. doi:10.1186/1476-511X-13-154.

[214] Terrence M Riley, Philip A Sapp, Penny M Kris-Etherton, and Kristina S Petersen. Effects of saturated fatty acid consumption on lipoprotein (a): a systematic review and meta-analysis of randomized controlled trials. *The American Journal of Clinical Nutrition*, 120(3):619–629, September 2024. doi:10.1016/j.ajcnut.2024.06.019.

[215] Konstantinos N. Aronis, Richard J. Joseph, George L. Blackburn, and Christos Mantzoros. trans-fatty acids, insulin resistance/diabetes, and cardiovascular disease risk: should policy decisions be based on observational cohort studies, or should we be waiting for results from randomized placebo-controlled trials? *Metabolism*, 60(7):901–905, July 2011. doi:10.1016/j.metabol.2011.04.003.

[216] Qi Sun. Trans fat and cardiovascular disease. *North American Journal of Medicine and Science*, 1(1):34–35, 2008. URL: https://najms.com/index.php/najms/article/view/397.

[217] I A Brouwer, A J Wanders, and M B Katan. Trans fatty acids and cardiovascular health: research completed? *European Journal of Clinical Nutrition*, 67(5):541–547, March 2013. doi:10.1038/ejcn.2013.43.

[218] Robert Clarke and Sarah Lewington. Trans fatty acids and coronary heart disease. *BMJ*, 333(7561):214, July 2006. doi:10.1136/bmj.333.7561.214.

[219] Dariush Mozaffarian, Martijn B. Katan, Alberto Ascherio, Meir J. Stampfer, and Walter C. Willett. Trans fatty acids and cardiovascular disease. *New England Journal of Medicine*, 354(15):1601–1613, April 2006. doi:10.1056/nejmra054035.

[220] A D McClain, W van den Bos, D Matheson, M Desai, S M McClure, and T N Robinson. Visual illusions and plate design: the effects of plate rim widths and rim coloring on perceived food portion size. *International Journal of Obesity*, 38(5):657–662, September 2013. doi:10.1038/ijo.2013.169.

[221] Barbara J. Rolls, Liane S. Roe, Kitti H. Halverson, and Jennifer S. Meengs. Using a smaller plate did not reduce energy intake at meals. *Appetite*, 49(3):652–660, November 2007. doi:10.1016/j.appet.2007.04.005.

[222] E. Robinson, S. Nolan, C. Tudur-Smith, E. J. Boyland, J. A. Harrold, C. A. Hardman, and J. C. G. Halford. Will smaller plates lead to smaller waists? a systematic review and meta-analysis of the effect that experimental manipulation of dishware size has on energy consumption. *Obesity Reviews*, 15(10):812–821, July 2014. doi:10.1111/obr.12200.

[223] Brian Wansink, Koert van Ittersum, and James E. Painter. Ice cream illusions. *American Journal of Preventive Medicine*, 31(3):240–243, September 2006. doi: 10.1016/j.amepre.2006.04.003.

[224] M. Barbara E. Livingstone and L. Kirsty Pourshahidi. Portion size and obesity. *Advances in Nutrition*, 5(6):829–834, November 2014. doi:10.3945/an.114.007104.

[225] Barbara J Rolls, Erin L Morris, and Liane S Roe. Portion size of food affects energy intake in normal-weight and overweight men and women. *The American Journal of Clinical Nutrition*, 76(6):1207–1213, December 2002. doi:10.1093/ajcn/76.6.1207.

[226] Krista Casazza, Andrew Brown, Arne Astrup, Fredrik Bertz, Charles Baum, Michelle Bohan Brown, John Dawson, Nefertiti Durant, Gareth Dutton, David A. Fields, Kevin R. Fontaine, Steven Heymsfield, David Levitsky, Tapan Mehta, Nir Menachemi, P.K. Newby, Russell Pate, Hollie Raynor, Barbara J. Rolls, Bisakha Sen, et al. Weighing the evidence of common beliefs in obesity research. *Critical Reviews in Food Science and Nutrition*, 55(14):2014–2053, June 2015. doi:10.1080/10408398.2014.922044.

[227] Meghan L. Butryn, Suzanne Phelan, James O. Hill, and Rena R. Wing. Consistent self-monitoring of weight: a key component of successful weight loss maintenance. *Obesity (Silver Spring)*, 15(12):3091–3096, 2007. doi:10.1038/oby.2007.368.

[228] Rena R. Wing, Deborah F. Tate, Amy A. Gorin, Hollie A. Raynor, and Joseph L. Fava. A self-regulation program for maintenance of weight loss. *New England Journal of Medicine*, 355(15):1563–1571, 2006. doi:10.1056/NEJMoa061883.

[229] Jeffrey J. VanWormer, Jennifer A. Linde, Lisa J. Harnack, Steven D. Stovitz, and Robert W. Jeffery. Self-weighing frequency is associated with weight gain prevention over 2 years among working adults. *International Journal of Behavioral Medicine*, 19(3):351–358, 2012. doi:10.1007/s12529-011-9166-2.

[230] Arne Astrup and Stephan Rossner. Lessons from obesity management programmes: greater initial weight loss improves long-term maintenance. *Obesity Reviews*, 1(1):17–19, 2000. doi:10.1046/j.1467-789x.2000.00005.x.

[231] Lisa M. Nackers, Kathryn M. Ross, and Michael G. Perri. The association between rate of initial weight loss and long-term success in obesity treatment: Does slow and steady win the race? *International Journal of Behavioral Medicine*, 17(3):161–167, May 2010. doi:10.1007/s12529-010-9092-y.

[232] Z. Kmietowicz. Gradual weight loss is no better than rapid weight loss for long term weight control, study finds. *BMJ*, 349:g6267, October 2014. doi:10.1136/bmj.g6267.

[233] Damoon Ashtary-Larky, Matin Ghanavati, Nasrin Lamuchi-Deli, Seyedeh Arefeh Payami, Sara Alavi-Rad, Mehdi Boustaninejad, Reza Afrisham, Amir Abbasnezhad, and Meysam Alipour. Rapid weight loss vs. slow weight loss: Which is more effective on body composition and metabolic risk factors? *International Journal of Endocrinology and Metabolism*, 15(3):e13249, May 2017. doi:10.5812/ijem.13249.

[234] Hiroyasu Mori. Effect of timing of protein and carbohydrate intake after resistance exercise on nitrogen balance in trained and untrained young men. *Journal of Physiological Anthropology*, 33(24):1–7, 2014. doi:10.1186/1880-6805-33-24.

[235] Brad J. Schoenfeld, Alan A. Aragon, and James W. Krieger. The effect of protein timing on muscle strength and hypertrophy: a meta-analysis. *Journal of the International Society of Sports Nutrition*, 10:53, 2013. doi:10.1186/1550-2783-10-53.

[236] Brad J. Schoenfeld and Alan A. Aragon. How much protein can the body use in a single meal for muscle-building? implications for daily protein distribution. *Journal of the International Society of Sports Nutrition*, 15:10, 2018. doi:10.1186/s12970-018-0215-1.

[237] Robert W. Morton, Kate T. Murphy, Sarah R. McKellar, Brad J. Schoenfeld, Menno Henselmans, Eric Helms, Alan A. Aragon, Michaela C. Devries, Leanne Banfield, James W. Krieger, and Stuart M. Phillips. A systematic review, meta-analysis and meta-regression of the effect of protein supplementation on resistance training-induced gains in muscle mass and strength in healthy adults. *British Journal of Sports Medicine*, 52(6):376–384, 2018. doi:10.1136/bjsports-2017-097608.

[238] Alan A. Aragon and Brad J. Schoenfeld. Nutrient timing revisited: is there a post-exercise anabolic window? *Journal of the International Society of Sports Nutrition*, 10:5, 2013. doi:10.1186/1550-2783-10-5.

[239] Alice Bellicha, Marleen A. van Baak, Francesca Battista, and et al. Effect of exercise training on weight loss, body composition changes, and weight maintenance in adults with overweight or obesity: An overview of 12 systematic reviews and 149 studies. *Obesity Reviews*, 22(S4):e13256, 2021. doi:10.1111/obr.13256.

[240] L. Schwingshackl, S. Dias, B. Strasser, and G. Hoffmann. Impact of different training modalities on anthropometric and metabolic characteristics in overweight/obese subjects: a systematic review and network meta-analysis. *PLoS ONE*, 8(12):e82853, 2013. doi:10.1371/journal.pone.0082853.

[241] Jason M. Curtis, Jose Antonio, and Cassandra Evans. Resistance training versus cardiovascular training: which is better for fat loss?: A brief narrative review. *Journal for Sports Neuroscience*, 1(2):Article 20, 2025. URL: https://nsuworks.nova.edu/neurosports/vol1/iss2/20.

[242] Su Hang, Lan Xiaoyu, Wang Jue, Lu Yingli, and Zhang Li. Effects of resistance training and aerobic training on improving the composition of middle-aged adults with obesity in an interventional study. *Scientific Reports*, 15:33972, 2025. doi:10.1038/s41598-025-11076-w.

[243] Mohamed Ahmed Said, Mohamed Abdelmoneem, Abdullah Almaqhawi, and et al. Multidisciplinary approach to obesity: Aerobic or resistance physical exercise? *Journal of Exercise Science & Fitness*, 16(2):118–123, 2018. doi:10.1016/j.jesf.2018.11.001.

[244] W. Kyle Mitchell, John Williams, Philip Atherton, Mike Larvin, John Lund, and Marco Narici. Sarcopenia, dynapenia, and the impact of advancing age on human skeletal muscle size and strength; a quantitative review. *Frontiers in Physiology*, 3, 2012. doi:10.3389/fphys.2012.00260.

[245] Alfonso J Cruz-Jentoft, Gülistan Bahat, Jürgen Bauer, Yves Boirie, Olivier Bruyère, Tommy Cederholm, Cyrus Cooper, Francesco Landi, Yves Rolland, Avan Aihie Sayer, Stéphane M Schneider, Cornel C Sieber, Eva Topinkova, Maurits Vandewoude, Marjolein Visser, Mauro Zamboni, Ivan Bautmans, Jean-Pierre Baeyens, Matteo Cesari, Antonio Cherubini, John Kanis, Marcello Maggio, Finbarr Martin, Jean-Pierre Michel, Kaisu Pitkala, Jean-Yves Reginster, René Rizzoli, Dolores Sánchez-Rodríguez, and Jos Schols. Sarcopenia: revised european consensus on definition and diagnosis. *Age and Ageing*, 48(1):16–31, September 2018. doi:10.1093/ageing/afy169.

[246] Mark D. Peterson, Matthew R. Rhea, Ananda Sen, and Paul M. Gordon. Resistance exercise for muscular strength in older adults: A meta-analysis. *Ageing Research Reviews*, 9(3):226–237, July 2010. doi:10.1016/j.arr.2010.03.004.

[247] Mette Midttun, Karsten Overgaard, Bo Zerahn, Maria Pedersen, Anahita Rashid, Peter Busch Østergren, Tine Kolenda Paulin, Thea Winther Pødenphanth, Linda Katharina Karlsson, Eva Rosendahl, Anne-Mette Ragle, Anders Vinther, and Rune Skovgaard Rasmussen. Beneficial effects of exercise, testosterone, vitamin d, calcium and protein in older men—a randomized clinical trial. *Journal of Cachexia, Sarcopenia and Muscle*, 15(4):1451–1462, June 2024. doi:10.1002/jcsm.13498.

[248] Arny A. Ferrando, Melinda Sheffield-Moore, Catherine W. Yeckel, Charles Gilkison, Jie Jiang, Alison Achacosa, Steven A. Lieberman, Kevin Tipton, Robert R. Wolfe, and Randall J. Urban. Testosterone administration to older men improves muscle function: molecular and physiological mechanisms. *American Journal of Physiology-Endocrinology and Metabolism*, 282(3):E601–E607, March 2002. doi:10.1152/ajpendo.00362.2001.

[249] Thomas W. Storer, Shehzad Basaria, Tinna Traustadottir, S. Mitchell Harman, Karol Pencina, Zhuoying Li, Thomas G. Travison, Renee Miciek, Panayiotis Tsitouras, Kathleen Hally, Grace Huang, and Shalender Bhasin. Effects of testosterone supplementation for 3-years on muscle performance and physical function in older men. *The Journal of Clinical Endocrinology & Metabolism*, pages jc.2016–2771, October 2016. doi:10.1210/jc.2016-2771.

[250] Myung Jun Shin, Yun Kyung Jeon, and In Joo Kim. Testosterone and sarcopenia. *The World Journal of Men's Health*, 36(3):192, 2018. doi:10.5534/wjmh.180001.

[251] C.D. Filippou, C.P. Tsioufis, C.G. Thomopoulos, C.C. Mihas, K.S. Dimitriadis, L.I. Sotiropoulou, C.A. Chrysochoou, P.I. Nihoyannopoulos, and D.M. Tousoulis. Dietary approaches to stop hypertension (dash) diet and blood pressure reduction in adults with and without hypertension: A systematic review and meta-analysis of randomized controlled trials. *Advances in Nutrition*, 11(5):1150–1160, 2020. doi:10.1093/advances/nmaa041.

[252] X. Theodoridis, M. Chourdakis, L. Chrysoula, V. Chroni, I. Tirodimos, K. Dipla, E. Gkaliagkousi, and A. Triantafyllou. Adherence to the dash diet and risk of hypertension: A systematic review and meta-analysis. *Nutrients*, 15(14):3261, 2023. doi:10.3390/nu15143261.

[253] Christina Filippou, Fotis Tatakis, Dimitrios Polyzos, Eleni Manta, Costas Thomopoulos, Petros Nihoyannopoulos, Dimitrios Tousoulis, and Konstantinos Tsioufis. Overview of salt restriction in the dietary approaches to stop hypertension (dash)

and the mediterranean diet for blood pressure reduction. *Reviews in Cardiovascular Medicine*, 23(1):36, 2022. doi:10.31083/j.rcm2301036.

[254] N.A. Reisdorph, A.E. Hendricks, M. Tang, K.A. Doenges, R.M. Reisdorph, B.C. Tooker, K. Quinn, S.J. Borengasser, Y. Nkrumah-Elie, D.N. Frank, W.W. Campbell, and N.F. Krebs. Nutrimetabolomics reveals food-specific compounds in urine of adults consuming a dash-style diet. *Scientific Reports*, 10(1):1157, 2020. doi:10.1038/s41598-020-57979-8.

[255] B.E. Wickman, B. Enkhmaa, R. Ridberg, E. Romero, M. Cadeiras, F. Meyers, and F. Steinberg. Dietary management of heart failure: Dash diet and precision nutrition perspectives. *Nutrients*, 13(12):4424, 2021. doi:10.3390/nu13124424.

[256] Patrick J. Gray, Sean D. Conklin, Todor I. Todorov, and Sasha M. Kasko. Cooking rice in excess water reduces both arsenic and enriched vitamins in the cooked grain. *Food Additives & Contaminants: Part A*, page 1–8, November 2015. doi:10.1080/19440049.2015.1103906.

[257] Tasila Mwale, Mohammad Mahmudur Rahman, and Debapriya Mondal. Risk and benefit of different cooking methods on essential elements and arsenic in rice. *International Journal of Environmental Research and Public Health*, 15(6):1056, May 2018. doi:10.3390/ijerph15061056.

[258] Fred Brouns. Phytic acid and whole grains for health controversy. *Nutrients*, 14(1):25, December 2021. doi:10.3390/nu14010025.

[259] Azusa Hirakawa, Seiichiro Aoe, Shaw Watanabe, Takayoshi Hisada, Jun Mochizuki, Shoichi Mizuno, Tanji Hoshi, and Sayuri Kodama. The nested study on the intestinal microbiota in GENKI study with special reference to the effect of brown rice eating. *Journal of Obesity and Chronic Diseases*, 03(01), 2019. doi:10.17756/jocd.2019-022.

[260] Jiayue Yu, Bhavadharini Balaji, Maria Tinajero, Sarah Jarvis, Tauseef Khan, Sudha Vasudevan, Viren Ranawana, Amudha Poobalan, Shilpa Bhupathiraju, Qi Sun, Walter Willett, Frank B Hu, David J A Jenkins, Viswanathan Mohan, and Vasanti S Malik. White rice, brown rice and the risk of type 2 diabetes: a systematic review and meta-analysis. *BMJ Open*, 12(9):e065426, September 2022. doi:10.1136/bmjopen-2022-065426.

[261] Zeinab Khosravi, Amir Hadi, Helda Tutunchi, Mohammad Asghari-Jafarabadi, Fatemeh Naeinie, Neda Roshanravan, Alireza Ostadrahimi, and Abdulmnannan Fadel. The effects of butyrate supplementation on glycemic control, lipid profile, blood pressure, nitric oxide level and glutathione peroxidase activity in type 2 diabetic patients: A randomized triple -blind, placebo-controlled trial. *Clinical Nutrition ESPEN*, 49:79–85, June 2022. doi:10.1016/j.clnesp.2022.03.008.

[262] Ana Navas-Acien and Keeve E. Nachman. Public health responses to arsenic in rice and other foods. *JAMA Internal Medicine*, 173(15):1395, August 2013. doi:10.1001/jamainternmed.2013.6405.

[263] P. Ravenscroft. Predicting the global distribution of natural arsenic contamination of groundwater. In *Symposium on arsenic: the geography of a global problem, Royal Geographical Society, London*, 2007. URL: http://www.geog.cam.ac.uk/research/projects/arsenic/symposium/S1.2_P_Ravenscroft.pdf.

[264] J.D. Ayotte, L. Medalie, S.L. Qi, L.C. Backer, and B.T. Nolan. Estimating the high-arsenic domestic-well population in the conterminous united states. *Environmental Science & Technology*, 51(21):12443–12454, 2017. doi:10.1021/acs.est.7b02881.

[265] U.S. EPA. Arsenic occurrence in public drinking water supplies. Technical Report EPA-815-R-00-23, U.S. Environmental Protection Agency, 2000. URL: https://hero.epa.gov/hero/index.cfm/reference/download/reference_id/1373136.

[266] D. Wilson, C. Hooper, and X. Shi. Arsenic and lead in juice: apple, citrus, and apple-base. *Journal of Environmental Health*, 75(5):14–20, 2012. URL: https://pubmed.ncbi.nlm.nih.gov/23270108/.

[267] K.E. Nachman, P.A. Baron, G. Raber, K.A. Francesconi, A. Navas-Acien, and D.C. Love. Roxarsone, inorganic arsenic, and other arsenic species in chicken: a u.s.-based market basket sample. *Environmental Health Perspectives*, 121(7):818–824, 2013. doi:10.1289/ehp.1206245.

[268] U.S. FDA. Arsenic in rice: Full analytical results from rice/rice product sampling - september 2012, 2012. U.S. Food and Drug Administration. URL: http://www.fda.gov/Food/FoodborneIllnessContaminants/Metals/ucm319916.htm.

[269] J.Y. Chung, S.D. Yu, and Y.S. Hong. Environmental source of arsenic exposure. *Journal of Preventive Medicine and Public Health*, 47(5):253–257, 2014. doi:10.3961/jpmph.14.036.

[270] Stephanie M. Eick and Craig Steinmaus. Arsenic and obesity: a review of causation and interaction. *Current Environmental Health Reports*, 7(3):343–351, 2020. doi:10.1007/s40572-020-00288-z.

[271] T. Farkhondeh, S. Samarghandian, and M. Azimi-Nezhad. The role of arsenic in obesity and diabetes. *Journal of Cellular Physiology*, 234(8):12516–12529, 2019. doi:10.1002/jcp.28112.

[272] A. et al. Ahangarpour. Chronic exposure to arsenic and high fat diet additively induced cardiotoxicity in male mice. *Research in Pharmaceutical Sciences*, 13(1):47–56, 2018. doi:10.4103/1735-5362.220967.

[273] D.Y. Garciafigueroa, L.R. Klei, F. Ambrosio, and A. Barchowsky. Arsenic-stimulated lipolysis and adipose remodeling is mediated by g-protein-coupled receptors. *Toxicological Sciences*, 134(2):335–344, 2013. doi:10.1093/toxsci/kft108.

[274] E.J. Ditzel, T. Nguyen, P. Parker, and T.D. Camenisch. Effects of arsenite exposure during fetal development on energy metabolism and susceptibility to diet-induced fatty liver disease in male mice. *Environmental Health Perspectives*, 124(2):201–209, 2016. doi:10.1289/ehp.1409501.

[275] C.M. Carmean, A.G. Kirkley, M. Landeche, H. Ye, B. Chellan, and H. et al. Aldirawi. Arsenic exposure decreases adiposity during high-fat feeding. *Obesity (Silver Spring)*, 28(5):932–941, 2020. doi:10.1002/oby.22770.

[276] J. Bae, Y. Jang, H. Kim, K. Mahato, C. Schaecher, and I.M. et al. Kim. Arsenite exposure suppresses adipogenesis, mitochondrial biogenesis and thermogenesis via autophagy inhibition in brown adipose tissue. *Scientific Reports*, 9(1):14464, 2019. doi:10.1038/s41598-019-50965-9.

[277] M. Vahter. Methylation of inorganic arsenic in different mammalian species and population groups. *Science Progress*, 82(Pt 1):69–88, 1999. doi:10.1177/003685049908200104.

[278] Z. Drobná, F.S. Walton, A.W. Harmon, D.J. Thomas, and M. Stýblo. Interspecies differences in metabolism of arsenic by cultured primary hepatocytes. *Toxicology and Applied Pharmacology*, 245(1):47–56, 2010. doi:10.1016/j.taap.2010.01.015.

[279] R. et al. Grashow. Inverse association between toenail arsenic and body mass index in a population of welders. *Environmental Research*, 131:131–133, 2014. doi:10.1016/j.envres.2014.03.010.

[280] A.S. Ettinger, P. Bovet, J. Plange-Rhule, T.E. Forrester, E.V. Lambert, and N. et al. Lupoli. Distribution of metals exposure and associations with cardiometabolic risk factors in the "modeling the epidemiologic transition study". *Environmental Health*, 13:90, 2014. doi:10.1186/1476-069x-13-90.

[281] H.C. Lin, Y.K. Huang, H.S. Shiue, L.S. Chen, C.S. Choy, and S.R. et al. Huang. Arsenic methylation capacity and obesity are associated with insulin resistance in obese children and adolescents. *Food and Chemical Toxicology*, 74:60–67, 2014. doi:10.1016/j.fct.2014.08.018.

[282] C.T. Su, H.C. Lin, C.S. Choy, Y.K. Huang, S.R. Huang, and Y.M. Hsueh. The relationship between obesity, insulin and arsenic methylation capability in taiwan adolescents. *Science of the Total Environment*, 414:152–158, 2012. doi:10.1016/j.scitotenv.2011.10.023.

[283] C.M. Bulka, S.L. Mabila, J.P. Lash, M.E. Turyk, and M. Argos. Arsenic and obesity: A comparison of urine dilution adjustment methods. *Environmental Health Perspectives*, 125(8):087020, 2017. doi:10.1289/ehp1202.

[284] H.S. Bae, D.Y. Ryu, B.S. Choi, and J.D. Park. Urinary arsenic concentrations and their associated factors in korean adults. *Toxicological Research*, 29(2):137–142, 2013. doi:10.5487/tr.2013.29.2.137.

[285] G. Velmurugan, K. Swaminathan, G. Veerasekar, J.Q. Purnell, S. Mohanraj, and M. et al. Dhivakar. Metals in urine in relation to the prevalence of pre-diabetes, diabetes and atherosclerosis in rural india. *Occupational and Environmental Medicine*, 75(9):661–667, 2018. doi:10.1136/oemed-2018-104996.

[286] R.C. Lewis, J.D. Meeker, N. Basu, A.M. Gauthier, A. Cantoral, and A. et al. Mercado-García. Urinary metal concentrations among mothers and children in a mexico city birth cohort study. *International Journal of Hygiene and Environmental Health*, 221(4):609–615, 2018. doi:10.1016/j.ijheh.2018.04.005.

[287] C. Steinmaus, F. Castriota, C. Ferreccio, A.H. Smith, Y. Yuan, J. Liaw, and G. Marshall. Obesity and excess weight in early adulthood and high risks of arsenic-related cancer in later life. *Environmental Research*, 142:594–601, 2015. doi:10.1016/j.envres.2015.07.021.

[288] F. Castriota, J. Acevedo, C. Ferreccio, A.H. Smith, J. Liaw, M.T. Smith, and C. Steinmaus. Obesity and increased susceptibility to arsenic-related type 2 diabetes in northern chile. *Environmental Research*, 167:248–254, 2018. doi:10.1016/j.envres.2018.07.022.

[289] T.C. Sung, J.W. Huang, and H.R. Guo. Association between arsenic exposure and diabetes: A meta-analysis. *BioMed Research International*, 2015:368087, 2015. doi:10.1155/2015/368087.

[290] A. Navas-Acien, E.K. Silbergeld, R.A. Streeter, J.M. Clark, T.A. Burke, and E. Guallar. Arsenic exposure and type 2 diabetes: a systematic review of the experimental and epidemiological evidence. *Environmental Health Perspectives*, 114(5):641–648, 2006. doi:10.1289/ehp.8551.

[291] C.J. Chen, S.L. Wang, J.M. Chiou, C.H. Tseng, H.Y. Chiou, and Y.M. et al. Hsueh. Arsenic and diabetes and hypertension in human populations: a review. *Toxicology and Applied Pharmacology*, 222(3):298–304, 2007. doi:10.1016/j.taap.2006.12.032.

[292] W.C. Pan, W.J. Seow, M.L. Kile, E.B. Hoffman, Q. Quamruzzaman, M. Rahman, and et al. Association of low to moderate levels of arsenic exposure with risk of type 2 diabetes in bangladesh. *American Journal of Epidemiology*, 178(10):1563–1570, 2013. doi:10.1093/aje/kwt195.

[293] C.H. Tseng, T.Y. Tai, C.K. Chong, C.P. Tseng, M.S. Lai, and B.J. et al. Lin. Long-term arsenic exposure and incidence of non-insulin-dependent diabetes mellitus: a cohort study in arseniasis-hyperendemic villages in taiwan. *Environmental Health Perspectives*, 108(9):847–851, 2000. doi:10.1289/ehp.00108847.

[294] A. Nardone, C. Ferreccio, J. Acevedo, W. Enanoria, A. Blair, A.H. Smith, and C. Steinmaus. The impact of bmi on non-malignant respiratory symptoms and lung function in arsenic exposed adults of northern chile. *Environmental Research*, 158:710–719, 2017. doi:10.1016/j.envres.2017.06.024.

[295] F. Wu, F. Jasmine, M.G. Kibriya, M. Liu, O. Wojcik, and F. et al. Parvez. Association between arsenic exposure from drinking water and plasma levels of cardiovascular markers. *American Journal of Epidemiology*, 175(12):1252–1261, 2012. doi:10.1093/aje/kwr464.

[296] E.M. Hall, J. Acevedo, F.G. Lopez, S. Cortes, C. Ferreccio, and A.H. et al. Smith. Hypertension among adults exposed to drinking water arsenic in northern chile. *Environmental Research*, 153:99–105, 2017. doi:10.1016/j.envres.2016.11.016.

[297] M.C. Huang, C. Douillet, E.N. Dover, C. Zhang, R. Beck, and A. et al. Tejan-Sie. Metabolic phenotype of wild-type and as3mt-knockout c57bl/6j mice exposed to inorganic arsenic: The role of dietary fat and folate intake. *Environmental Health Perspectives*, 126(12):127003, 2018. doi:10.1289/ehp3951.

[298] D.S. Paul, F.S. Walton, R.J. Saunders, and M. Stýblo. Characterization of the impaired glucose homeostasis produced in c57bl/6 mice by chronic exposure to arsenic and high-fat diet. *Environmental Health Perspectives*, 119(8):1104–1109, 2011. doi:10.1289/ehp.1003324.

[299] M. Tan, R.H. Schmidt, J.I. Beier, W.H. Watson, H. Zhong, and J.C. et al. States. Chronic subhepatotoxic exposure to arsenic enhances hepatic injury caused by high fat diet in mice. *Toxicology and Applied Pharmacology*, 257(3):356–364, 2011. doi:10.1016/j.taap.2011.09.019.

[300] Y. Zhang, J.L. Young, L. Cai, Y.G. Tong, L. Miao, and J.H. Freedman. Chronic exposure to arsenic and high fat diet induces sex-dependent pathogenic effects on

the kidney. *Chemico-Biological Interactions*, 310:108719, 2019. doi:10.1016/j.cbi.2019.06.032.

[301] Jun Tang, Ju-Sheng Zheng, Ling Fang, Yongxin Jin, Wenwen Cai, and Duo Li. Tea consumption and mortality of all cancers, cvd and all causes: a meta-analysis of eighteen prospective cohort studies. *British Journal of Nutrition*, 114(5):673–683, July 2015. doi:10.1017/s0007114515002329.

[302] R. Poole, O.J. Kennedy, P. Roderick, J.A. Fallowfield, P.C. Hayes, and J. Parkes. Coffee consumption and health: umbrella review of meta-analyses of multiple health outcomes. *BMJ*, 359:j5024, 2017. doi:10.1136/bmj.j5024.

[303] M. Carlström and S.C. Larsson. Coffee consumption and reduced risk of developing type 2 diabetes: a systematic review with meta-analysis. *Nutr Rev*, 76(6):395–417, 2018. doi:10.1093/nutrit/nuy014.

[304] Wan-Shui Yang, Wei-Ye Wang, Wen-Yan Fan, Qin Deng, and Xin Wang. Tea consumption and risk of type 2 diabetes: a dose–response meta-analysis of cohort studies. *British Journal of Nutrition*, 111(8):1329–1339, December 2013. doi:10.1017/s0007114513003887.

[305] M. Ebadi, S. Ip, R.A. Bhanji, and A.J. Montano-Loza. Effect of coffee consumption on non-alcoholic fatty liver disease incidence, prevalence and risk of significant liver fibrosis: systematic review with meta-analysis of observational studies. *Nutrients*, 13(9):3042, 2021. doi:10.3390/nu13093042.

[306] Umar Hayat, Ali A. Siddiqui, Hayrettin Okut, Saba Afroz, Syed Tasleem, and Ahmed Haris. The effect of coffee consumption on the non-alcoholic fatty liver disease and liver fibrosis: A meta-analysis of 11 epidemiological studies. *Annals of Hepatology*, 20:100254, January 2021. doi:10.1016/j.aohep.2020.08.071.

[307] Stanisław Surma, Amirhossein Sahebkar, and Maciej Banach. Coffee or tea: Anti-inflammatory properties in the context of atherosclerotic cardiovascular disease prevention. *Pharmacological Research*, 187:106596, 2023. doi:10.1016/j.phrs.2022.106596.

[308] Yi Zhang and Dian-Zhong Zhang. Associations of coffee consumption with circulating level of adiponectin and leptin. a meta-analysis of observational studies. *International Journal of Food Sciences and Nutrition*, 69(8):1003–1012, March 2018. doi:10.1080/09637486.2018.1445202.

[309] M. Mahdavi-Roshan, A. Salari, Z. Ghorbani, and A. Ashouri. The effects of regular consumption of green or black tea beverage on blood pressure in those with elevated blood pressure or hypertension: a systematic review and meta-analysis. *Complement Ther Med*, 51:102430, 2020. doi:10.1016/j.ctim.2020.102430.

[310] Renfan Xu, Ke Yang, Sui Li, Meiyan Dai, and Guangzhi Chen. Effect of green tea consumption on blood lipids: a systematic review and meta-analysis of randomized controlled trials. *Nutrition Journal*, 19(1), September 2020. doi:10.1186/s12937-020-00557-5.

[311] Marilyn C Cornelis and Rob M van Dam. Habitual coffee and tea consumption and cardiometabolic biomarkers in the uk biobank: The role of beverage types and genetic variation. *The Journal of Nutrition*, 150(10):2772–2788, October 2020. doi:10.1093/jn/nxaa212.

[312] P. Chatterjee, S. Chandra, P. Dey, and S. Bhattacharya. Evaluation of anti-inflammatory effects of green tea and black tea: a comparative in vitro study. *J Adv Pharm Technol Res*, 3(2):136–138, 2012. doi:10.4103/2231-4040.97298.

[313] Laura O'Connor, Fumiaki Imamura, Soren Brage, Simon J. Griffin, Nicholas J. Wareham, and Nita G. Forouhi. Intakes and sources of dietary sugars and their association with metabolic and inflammatory markers. *Clinical Nutrition*, 37(4):1313–1322, August 2018. doi:10.1016/j.clnu.2017.05.030.

[314] Kye-Yeung Park, Hoon-Ki Park, and Hwan-sik Hwang. Relationship between abdominal obesity and alcohol drinking pattern in normal-weight, middle-aged adults: the korea national health and nutrition examination survey 2008–2013. *Public Health Nutrition*, 20(12):2192–2200, June 2017. doi:10.1017/s1368980017001045.

[315] Ulf Risérus and Erik Ingelsson. Alcohol intake, insulin resistance, and abdominal obesity in elderly men. *Obesity*, 15(7):1766–1773, July 2007. doi:10.1038/oby.2007.210.

[316] Joan M. Dorn, Kathleen Hovey, Paola Muti, Jo L. Freudenheim, Maurizio Trevisan, Marcia Russell, and Thomas H. Nochajski. Alcohol drinking patterns differentially affect central adiposity as measured by abdominal height in women and men. *The Journal of Nutrition*, 133(8):2655–2662, August 2003. doi:10.1093/jn/133.8.2655.

[317] Suk Hwa Jung, Kyoung Hwa Ha, and Dae Jung Kim. Visceral fat mass has stronger associations with diabetes and prediabetes than other anthropometric obesity indicators among korean adults. *Yonsei Medical Journal*, 57(3):674, 2016. doi:10.3349/ymj.2016.57.3.674.

[318] Osama Hamdy, Sriurai Porramatikul, and Ebaa Al-Ozairi. Metabolic obesity: the paradox between visceral and subcutaneous fat. *Current Diabetes Reviews*, 2(4):367–373, 2006. doi:10.2174/1573399810602040367.

[319] Ylenia Duca, Antonio Aversa, Rosita Angela Condorelli, Aldo Eugenio Calogero, and Sandro La Vignera. Substance abuse and male hypogonadism. *Journal of Clinical Medicine*, 8(5):732, May 2019. doi:10.3390/jcm8050732.

[320] Tina Kold Jensen, Mads Gottschau, Jens Otto Broby Madsen, Anne-Maria Andersson, Tina Harmer Lassen, Niels E Skakkebæk, Shanna H Swan, Lærke Priskorn, Anders Juul, and Niels Jørgensen. Habitual alcohol consumption associated with reduced semen quality and changes in reproductive hormones; a cross-sectional study among 1221 young danish men. *BMJ Open*, 4(9):e005462, September 2014. doi:10.1136/bmjopen-2014-005462.

[321] Belinda S Lennerz, Jacob T Mey, Owen H Henn, and David S Ludwig. Behavioral characteristics and self-reported health status among 2029 adults consuming a "carnivore diet". *Current Developments in Nutrition*, 5(12):nzab133, December 2021. doi:10.1093/cdn/nzab133.

[322] Leon Caly, Julian D. Druce, Mike G. Catton, David A. Jans, and Kylie M. Wagstaff. The fda-approved drug ivermectin inhibits the replication of sars-cov-2 in vitro. *Antiviral Research*, 178:104787, June 2020. doi:10.1016/j.antiviral.2020.104787.

[323] Aleksandra Barac, Michele Bartoletti, Ozlem Azap, Linda Bussini, Onder Ergonul, Robert Krause, José Ramón Paño-Pardo, Nicholas R. Power, Jesús Rodríguez-Baño, Marcella Sibani, Balint Gergely Szabo, Sotirios Tsiodras, Paul E. Verweij,

Alejandro Martín Quirós, and Ines Zollner-Schwetz. Inappropriate use of ivermectin during the covid-19 pandemic: primum non nocere! *Clinical Microbiology and Infection*, 28(7):908–910, July 2022. `doi:10.1016/j.cmi.2022.03.022`.

[324] Farah Yasmin, Muhammad Sohaib Asghar, Unaiza Naeem, Hala Najeeb, Hamza Nauman, Muhammad Nadeem Ahsan, and Abdullah Khan Khattak. Self-medication practices in medical students during the covid-19 pandemic: A cross-sectional analysis. *Frontiers in Public Health*, 10, March 2022. `doi: 10.3389/fpubh.2022.803937`.

[325] Mark Bailey. A new equilibrium? economy and society, 1375 to 1400. In *After the Black Death*, pages 234–282. Oxford University PressOxford, February 2021.

[326] Christopher F. Bosio, Clayton O. Jarrett, Dana P. Scott, Jonathan Fintzi, and B. Joseph Hinnebusch. Comparison of the transmission efficiency and plague progression dynamics associated with two mechanisms by which fleas transmit yersinia pestis. *PLOS Pathogens*, 16(12):e1009092, December 2020. `doi:10.1371/journal.ppat.1009092`.

[327] Norman F Cantor. *In the wake of the plague*. HarperCollins, New York, NY, April 2002.

[328] Donald McNamara. The fifty year rehabilitation of the egg. *Nutrients*, 7(10):8716–8722, October 2015. `doi:10.3390/nu7105429`.

[329] Cristin E. Kearns, Laura A. Schmidt, and Stanton A. Glantz. Sugar industry and coronary heart disease research: A historical analysis of internal industry documents. *JAMA Internal Medicine*, 176(11):1680, November 2016. `doi:10.1001/jamainternmed.2016.5394`.

[330] Krista S. Crider, Yan Ping Qi, Lorraine F. Yeung, Cara T. Mai, Lauren Head Zauche, Arick Wang, Kelicia Daniels, and Jennifer L. Williams. Folic acid and the prevention of birth defects: 30 years of opportunity and controversies. *Annual Review of Nutrition*, 42(1):423–452, August 2022. `doi:10.1146/annurev-nutr-043020-091647`.

[331] Tylor J. Cosgrove and Christopher P. Murphy. Narcissistic susceptibility to conspiracy beliefs exaggerated by education, reduced by cognitive reflection. *Frontiers in Psychology*, 14, July 2023. `doi:10.3389/fpsyg.2023.1164725`.

[332] Lukasz Stasielowicz. Who believes in conspiracy theories? a meta-analysis on personality correlates. *Journal of Research in Personality*, 98:104229, June 2022. `doi:10.1016/j.jrp.2022.104229`.

[333] Shauna M. Bowes, Thomas H. Costello, and Arber Tasimi. The conspiratorial mind: A meta-analytic review of motivational and personological correlates. *Psychological Bulletin*, 149(5–6):259–293, May 2023. `doi:10.1037/bul0000392`.

[334] Briony D. Pulford, Andrew M. Colman, Eike K. Buabang, and Eva M. Krockow. The persuasive power of knowledge: Testing the confidence heuristic. *Journal of Experimental Psychology: General*, 147(10):1431–1444, October 2018. `doi:10.1037/xge0000471`.

[335] Shane Littrell, Jonathan A. Fugelsang, and Evan F. Risko. The metacognitive abilities of narcissists: Individual differences between grandiose and vulnerable subtypes. *Personality and Individual Differences*, 221:112570, April 2024. `doi:10.1016/j.paid.2024.112570`.

[336] Keith E. Stanovich, Richard F. West, and Maggie E. Toplak. Myside bias, rational thinking, and intelligence. *Current Directions in Psychological Science*, 22(4):259–264, August 2013. doi:10.1177/0963721413480174.

[337] Milan Toma. *The Economy of Distrust: Medical Misinformation and the Business of Conspiracy*. Dawning Research Press, New York, NY, 2025. URL: https://openlibrary.org/works/OL44048120W.

[338] Sahdeo Prasad, Subash C. Gupta, Amit K. Tyagi, and Bharat B. Aggarwal. Curcumin, a component of golden spice: From bedside to bench and back. *Biotechnology Advances*, 32(6):1053–1064, November 2014. doi:10.1016/j.biotechadv.2014.04.004.

[339] Binbin Chen, Lanqiu Tao, Min Tian, and Zhaohua Ji. Efficacy and safety of combination of semaglutide and basal insulin in patients with of type 2 diabetes mellitus: A systematic review and meta-analysis. *Clinical Nutrition ESPEN*, January 2025. doi:10.1016/j.clnesp.2025.01.056.

[340] John P.H. Wilding, Rachel L. Batterham, Salvatore Calanna, Melanie Davies, Luc F. Van Gaal, Ildiko Lingvay, Barbara M. McGowan, Julio Rosenstock, Marie T.D. Tran, Thomas A. Wadden, Sean Wharton, Koutaro Yokote, Niels Zeuthen, and Robert F. Kushner. Once-weekly semaglutide in adults with overweight or obesity. *New England Journal of Medicine*, 384(11):989–1002, March 2021. doi:10.1056/nejmoa2032183.

[341] Catalin Feier, Razvan Vonica, Alaviana Faur, Diana Streinu, and Calin Muntean. Assessment of thyroid carcinogenic risk and safety profile of glp1-ra semaglutide (ozempic) therapy for diabetes mellitus and obesity: A systematic literature review. *International Journal of Molecular Sciences*, 25(8):4346, April 2024. doi:10.3390/ijms25084346.

[342] Philip R Schauer and Amy E Rothberg. Point-counterpoint debate: Surgery vs medical treatment for the management of obesity. *The Journal of Clinical Endocrinology & Metabolism*, December 2024. doi:10.1210/clinem/dgae888.

[343] Kevin Aslett, Zeve Sanderson, William Godel, Nathaniel Persily, Jonathan Nagler, and Joshua A. Tucker. Online searches to evaluate misinformation can increase its perceived veracity. *Nature*, 625(7995):548–556, December 2023. doi:10.1038/s41586-023-06883-y.

[344] Audrey Eichenberger, Stephen Thielke, and Adam Van Buskirk. A case of bromism influenced by use of artificial intelligence. *Annals of Internal Medicine: Clinical Cases*, 4(8), August 2025. doi:10.7326/aimcc.2024.1260.

Dawning Research Press

Founded in 2019, Dawning Research Press is dedicated to publishing scholarly works that bridge academic rigor with public understanding. Our mission is to illuminate complex issues at the intersection of science, society, and human experience.

Contact Information

Email: admin@dawningresearch.org
Website: www.dawningresearch.org

For information about permissions, bulk purchases, educational discounts, or media inquiries, please contact us at the email address above.

Committed to evidence-based scholarship and accessible science communication.